THE PHYSICS OF STAR TREK

LAWRENCE M. KRAUSS

WITH A FOREWORD BY
STEPHEN HAWKING

A Member of the Perseus Books Group
New York

Published by Basic Books
A Member of the Perseus Books Group

Books published by Basic Books are available at special discounts for bulk purchases in the United States by corporations, institutions, and other organizations. For more information, please contact the Special Markets Department at the Perseus Books Group, 2300 Chestnut Street, Philadelphia, PA 19103 or e-mail special.markets@perseusbooks.com.

Designed by Brent Wilcox

A CIP record for this book is available from the Library of Congress.
ISBN-13: 978-0-465-00204-7
ISBN-10: 0-465-00204-8

10 9 8 7 6

To my family

"But I canna change the laws of physics, Captain!"

(Scotty, to Kirk, innumerable times)

CONTENTS

SECTION TWO
MATTER MATTER EVERYWHERE

In which the reader explores transporter beams, warp drives, dilithium crystals, matter-antimatter engines, and the holodeck

SECTION THREE
THE INVISIBLE UNIVERSE, OR THINGS THAT GO BUMP IN THE NIGHT

In which we speak of things that may exist but are not yet seen—extraterrestrial life, multiple dimensions, and an exotic zoo of other physics possibilities and impossibilities

FOREWORD

Stephen Hawking

I was very pleased that Data decided to call Newton, Einstein, and me for a game of poker aboard the *Enterprise*. Here was my chance to turn the tables on the two great men of gravity, particularly Einstein, who didn't believe in chance or in God playing dice. Unfortunately, I never collected my winnings because the game had to be abandoned on account of a red alert. I contacted Paramount studios afterward to cash in my chips, but they didn't know the exchange rate.

Science fiction like Star Trek is not only good fun but it also serves a serious purpose, that of expanding the human imagination. We may not yet be able to boldly go where no man (or woman) has gone before, but at least we can do it in the mind. We can explore how the human spirit might respond to future developments in science and we can speculate on what those developments might be. There is a two-way trade between science fiction and science. Science fiction suggests ideas that scientists incorporate into their theories, but sometimes science turns up notions that are stranger than any science fiction. Black holes are an example, greatly assisted by the inspired name that the physicist John Archibald Wheeler gave them. Had they continued with their original names of "frozen stars" or "gravitationally completely collapsed objects," there wouldn't have been half so much written about them.

One thing that Star Trek and other science fiction have focused attention on is travel faster than light. Indeed, it is absolutely essential to Star Trek's story line. If the *Enterprise* were restricted to flying just under the speed of light, it might seem to the crew that the round trip to the center of the galaxy took only a few years, but 80,000 years would have elapsed on Earth before the spaceship's return. So much for going back to see your family!

Fortunately, Einstein's general theory of relativity allows the possibility for a way around this difficulty: one might be able to warp spacetime and create a shortcut between the places one wanted to visit. Although there are problems of negative energy, it seems that such warping might be within our capabilities in the future. There has not been much serious scientific research along these lines, however, partly, I think, because it sounds too much like science fiction. One of the consequences of rapid interstellar travel would be that one could also travel back in time. Imagine the outcry about the waste of taxpayers' money if it were known that the National Science Foundation were supporting research on time travel. For this reason, scientists working in this field have to disguise their real interest by using technical terms like "closed timelike curves" that are code for time travel. Nevertheless, today's science fiction is often tomorrow's science fact. The physics that underlies Star Trek is surely worth investigating. To confine our attention to terrestrial matters would be to limit the human spirit.

PREFACE

Why the physics of Star Trek? Gene Roddenberry's creation is, after all, science fiction, not science fact. Many of the technical wonders in the series therefore inevitably rest on notions that may be ill defined or otherwise at odds with our current understanding of the universe. I did not want to write a book that ended up merely outlining where the Star Trek writers went wrong.

Yet I found that I could not get the idea of this book out of my head. I confess that it was really the transporter that seduced me. Thinking about the challenges that would have to be faced in devising such a fictional technology forces one to ponder topics ranging from computers and the information superhighway to particle physics, quantum mechanics, nuclear energy, telescope building, biological complexity, and even the possible existence of the human soul! Compound this with ideas such as warped space and time travel and the whole subject became irresistible.

I soon realized that what made this so fascinating to me was akin to what keeps drawing fans to Star Trek today, almost thirty years after the series first aired. This is, as the omnipotent Star Trek prankster Q put it, "charting the unknown possibilities of existence." And, as I am sure Q would have agreed, it is even good fun to imagine them.

As Stephen Hawking states in the foreword to this book, science fiction like Star Trek helps expand the human imagination. Indeed, exploring the infinite possibilities the future holds—including a world where humanity has overcome its myopic international and racial tensions and ventured out to explore the universe in peace—is part of the continuing wonder of Star Trek. And, as I see this as central to the continuing wonder of modern physics, it is these possibilities that I have chosen to concentrate on here.

Based on an informal survey I carried out while walking around my university campus the other day, the number of people in the United States who would not recognize the phrase "Beam me up, Scotty" is roughly comparable to the number of people who have never heard of ketchup. When we consider that the Smithsonian Institution's exhibition on the starship *Enterprise* was the most popular display in their Air and Space Museum—more popular than the real spacecraft there—I think it is clear that Star Trek is a natural vehicle for many people's curiosity about the universe. What better context to introduce some of the more remarkable ideas at the forefront of today's physics and the threshold of tomorrow's? I hope you find the ride as enjoyable as I have.

Live long and prosper.

PREFACE TO THE REVISED EDITION

When I first sat down to write *The Physics of Star Trek* almost 13 years ago I had no idea how significantly it would change my life, nor of the impact it might have on trekkers and non-trekkers alike. I was mostly hoping that following its publication a mob of angry fans wouldn't lynch me and that my physics colleagues would still talk to me.

Needless to say, these worries proved to be ill-founded. Indeed the immediate and overwhelming reaction on all counts was the opposite of what I had expected. One of the first letters I received after the book appeared was from a fan who said, "I had been waiting for 20 years to read a Star Trek book in the Science Fact section of a bookstore!" And when I began to lecture on this subject I met 7- and 8-year-olds with dog-eared copies of the book who had great questions to ask. And my colleagues turned out to be largely thrilled that a physics book could actually be a popular bestseller. And lo and behold, the book appeared to create a new genre of "The Science of . . ." books. First, books titled *The ______ of Star Trek* began to appear by the dozens, followed quickly by books with titles like *The Physics of Christmas* and *The Science of Harry Potter.*

And I even got to stand at the helm of the *Enterprise* at Paramount, even if I didn't get to join a poker game with Einstein, and I filmed a TV documentary with Captain Kirk himself, and hung out with the likes of Commander Riker and Quark.

Shortly after the book appeared I was asked for a sequel, and the request has been repeated numerous times over the years, but I decided I had said everything I wanted to say on this subject. Well, almost everything I had to say. In the intervening years, not only has Star Trek continued, but the world of science has as well, and I daresay the latter may have progressed far more than the former. In an effort to bring the science in the book up to date I decided to review the material from cover to cover, adding new information when necessary, and removing arguments when nature has shown them to be incorrect.

Of course, in the process, I couldn't resist adding some new Star Trek connections and even a few new bloopers, one related to me by a 5-year-old at a lecture I gave, and one by a member of the crew of the *Enterprise.* I have tried hard to preserve the character of the original book, much of which has happily survived unscathed. In the end, I hope readers continue to enjoy the discussions and come away a bit more enamored with the amazing fact that, as remarkable as the Star Trek Universe may be, the real universe keeps providing surprises that are both grander and stranger than anything human screenwriters may come up with.

Lawrence M. Krauss
Cleveland, Ohio 2007

SECTION ONE

A COSMIC POKER GAME

In which the physics of inertial dampers and tractor beams paves the way for time travel, warp speed, deflector shields, wormholes, and other spacetime oddities

1

Newton Antes

"No matter where you go, there you are."

—From a plaque on the starship *Excelsior*, in *Star Trek VI: The Undiscovered Country*, presumably borrowed from *The Adventures of Buckaroo Banzai*

You are at the helm of the starship *Defiant* (*NCC–1764*), currently in orbit around the planet Iconia, near the Neutral Zone. Your mission: to rendezvous with a nearby supply vessel at the other end of this solar system in order to pick up components to repair faulty transporter primary energizing coils. There is no need to achieve warp speeds; you direct the impulse drive to be set at full power for leisurely half-light-speed travel, which should bring you to your destination in a few hours, giving you time to bring the captain's log up to date. However, as you begin to pull out of orbit, you feel an intense pressure in your chest. Your hands are leaden, and you are glued to your seat. Your mouth is fixed in an evil-looking grimace, your eyes feel like they are about to burst out of their sockets, and the blood flowing through your body refuses to rise to your head. Slowly, you lose consciousness . . . and within minutes you die.

What happened? It is not the first signs of spatial "interphase" drift, which will later overwhelm the ship, or an attack from a previously cloaked Romulan vessel. Rather, you have fallen prey to something far more powerful. The ingenious writers of Star Trek, on whom you depend, have not yet invented inertial dampers, which they will introduce sometime later in the series. You have been defeated by nothing more exotic than Isaac Newton's laws of motion—the very first things one can forget about high school physics.

OK, I know some trekkers out there are saying to themselves, "How lame! Don't give me Newton. Tell me things I really want to know, like 'How does warp drive work?' or 'What is the flash before going to warp speed—is it like a sonic boom?' or 'What is a dilithium crystal anyway?'" All I can say is that we will get there eventually. Travel in the Star Trek universe involves some of the most exotic concepts in physics. But many different aspects come together before we can really address everyone's most fundamental question about Star Trek: "Is any of this *really* possible, and if so, *how*?"

To go where no one has gone before—indeed, before we even get out of Starfleet Headquarters—we first have to confront the same peculiarities that Galileo and Newton did over three hundred years ago. The ultimate motivation will be the truly cosmic question which was at the heart of

Gene Roddenberry's vision of Star Trek and which, to me, makes this whole subject worth thinking about: "*What does modern science allow us to imagine about our possible future as a civilization?*"

Anyone who has ever been in an airplane or a fast car knows the feeling of being pushed back into the seat as the vehicle accelerates from a standstill. This phenomenon works with a vengeance aboard a starship. The fusion reactions in the impulse drive produce huge pressures, which push gases and radiation backward away from the ship at high velocity. It is the backreaction force on the engines—from the escaping gas and radiation—that causes the engines to "recoil" forward. The ship, being anchored to the engines, also recoils forward. At the helm, you are pushed forward too, by the force of the captain's seat on your body. In turn, your body pushes back on the seat.

Now, here's the catch. Just as a hammer driven at high velocity toward your head will produce a force on your skull which can easily be lethal, the captain's seat will kill you if the force it applies to you is too great. Jet pilots and NASA have a name for the force exerted on your body while you undergo high accelerations (as in a plane or during a space launch): G-forces. I can describe these by recourse to my aching back: As I am sitting at my computer terminal busily typing, I feel the ever-present pressure of my office chair on my buttocks—a pressure that I have learned to live with (yet, I might add, that my buttocks are slowly reacting to in a very noncosmetic way). The force on my buttocks results from the pull of gravity, which if given free rein would accelerate me downward into the Earth. What stops me from accelerating—indeed, from moving beyond my seat—is the ground exerting an opposite upward force on my house's concrete and steel frame, which exerts an upward force on the wood floor of my second-floor study, which exerts a force on my chair, which in turn exerts a force on the part of my body in contact with it. If the Earth were twice as massive but had the same diameter, the pressure on my buttocks would be twice as great. The upward forces would have to compensate for the force of gravity by being twice as strong.

The same factors must be taken into account in space travel. If you are in the captain's seat and you issue a command for the ship to accelerate, you must take into account the force with which the seat will push you forward. If you request an acceleration twice as great, the force on you from the seat will be twice as great. The greater the acceleration, the greater the push. The only problem is that nothing can withstand the kind of force needed to accelerate to impulse speed quickly—certainly not your body.

By the way, this same problem crops up in different contexts throughout Star Trek—even on Earth. At the beginning of *Star Trek V: The Final Frontier,* James Kirk is free-climbing while on vacation in Yosemite when he slips and falls. Spock, who has on his rocket boots, speeds to the rescue, aborting the captain's fall within a foot or two of the ground. Unfortunately, this is a case where the solution can be as bad as the problem. It is the process of stopping over a distance of a few inches that can kill you, whether or not it is the ground that does the stopping or Spock's Vulcan grip.

Well before the reaction forces that will physically tear or break your body occur, other severe physiological problems set in. First and foremost, it becomes impossible for your heart to pump strongly enough to force the blood up to your head. This is why fighter pilots sometimes black out when they perform maneuvers involving rapid acceleration. Special suits have been created to force the blood up from pilots' legs to keep them conscious during acceleration. This physiological reaction remains one of the limiting factors in determining how fast the acceleration of present-day spacecraft can be, and it is why NASA, unlike Jules Verne in his classic *From the Earth to the Moon,* has never launched three men into orbit from a giant cannon.

If I want to accelerate from rest to, say, 150,000 km/sec, or about half the speed of light, I have to do it gradually, so that my body will not be torn apart in the process. In order not to be pushed back into my seat with a force greater than 3G, my acceleration must be no more than three times the downward acceleration of falling objects on Earth. At this rate of acceleration, it would take some 5 million seconds, or about 2 1/2 months, to reach half light speed! This would not make for an exciting episode.

To resolve this dilemma, sometime after the production of the first Constitution Class starship—the *Enterprise* (*NCC–1701*)—the Star Trek writers had to develop a response to the criticism that the accelerations aboard a starship would instantly turn the crew into "chunky salsa."[1] They came up with "inertial dampers," a kind of cosmic shock absorber and an ingenious plot device designed to get around this sticky little problem. (In fact, a full century earlier, the NX Class starship *Enterprise* commanded by Captain Archer seems to have been fitted with inertial dampers. Alas, while this ship was built much earlier, it was created by the writers much later.)

The inertial dampers are most notable in their absence. For example, the *Enterprise* was nearly destroyed after losing control of the inertial dampers when the microchip life-forms known as Nanites, as part of their evolutionary process, started munching on the ship's central-computer-core memory. Indeed, almost every time the *Enterprise* is destroyed (usually in some renegade timeline), the destruction is preceded by loss of the inertial dampers. The results of a similar loss of control in a Romulan Warbird provided us with an explicit demonstration that Romulans bleed green.

Alas, as with much of the technology in the Star Trek universe, it is much easier to describe the problem the inertial dampers address than it is to explain exactly how they might do it. The First Law of Star Trek physics surely must state that the more basic the problem to be circumvented, the more challenging the required solution must be. The reason we have come this far, and the reason we can even postulate a Star Trek future, is that physics is a field that builds on itself. A Star Trek fix must circumvent not merely some problem in physics but every bit of physical knowledge that has been built upon this problem. Physics progresses not by revolutions, which do away with all that went before, but rather by evolutions, which exploit the best about what is already understood. Newton's laws will continue to be as true a million years from now as they are today, no matter what we discover at the frontiers of science. If we drop a ball on Earth, it will always fall. If I sit at this desk and write from here to eternity, my buttocks will always suffer the same consequences.

Be that as it may, it would be unfair simply to leave the inertial dampers hanging without at least some concrete description of how they would have to operate. From what I have argued, they must create an artificial world inside a starship in which the reaction force that responds to the accelerating force is canceled. The objects inside the ship are "tricked" into acting as though they were not accelerating. I have described how accelerating gives you the same feeling as being pulled at by gravity. This connection, which was the basis of Einstein's general theory of relativity, is much more intimate than it may at first seem. Thus there is only one choice for the modus operandi of these gadgets: they must set up an artificial gravitational field inside the ship which "pulls" in the opposite direction to the reaction force, thereby canceling it out.

Even if you buy such a possibility, other practical issues must be dealt with. For one thing, it takes some time for the inertial dampers to kick in when unexpected impulses arise. For example, when the *Enterprise* was bumped into a causality loop by the *Bozeman* as the latter vessel emerged from a temporal distortion, the crew was thrown all about the bridge (even before the breach in the warp core and the failure of the dampers). I have read in the *Enterprise*'s technical specifications that the response time for the inertial dampers is about 60 milliseconds.[2] Short as this may seem, it would be long enough to kill you if the same delay occurred during programmed periods of acceleration. To convince yourself, think how long it takes for a hammer to smash your head open, or how long it takes for the ground to kill you if you hit it after falling off of a cliff in Yosemite. Just remember that a collision at 10 miles per hour is equivalent to running full speed into a brick wall! The inertial dampers had better be pretty quick to respond. More than one trekker I know has remarked that whenever the ship *is* buffeted, no one ever gets thrown more than a few feet.

Before leaving the familiar world of classical physics, I can't help mentioning another technological marvel that must confront Newton's laws in order to operate: the *Enterprise*'s tractor beam—highlighted in the rescue of the Genome colony on Moab IV, when it deflected an approaching stellar core fragment, and in a similar (but failed) attempt to save Bre'el IV by pushing an asteroidal moon back into its orbit. On the face of it, the tractor beam seems simple enough—more or less like an invisible rope or rod—even if the force exerted may be exotic. Indeed, just like a strong rope, the tractor beam often does a fine job of pulling in a shuttle craft, towing another ship, or inhibiting the escape of an enemy spacecraft. In fact, before the Federation had universal access to tractor beams, the NX Class *Enterprise* apparently used a magnetic "grappler" for precisely these tasks. No matter; the only problem is that when we pull something, be it with a rope, a grappler, or a tractor beam, we must be anchored to the ground or to something else heavy. Anyone who has ever been skating knows what happens if you are on the ice and you try to push someone away from you. You do manage to separate, but at your own expense. Without any firm grounding, you are a helpless victim of your own inertia.

It was this very principle that prompted Captain Jean-Luc Picard to order Lieutenant Riker to turn off the tractor beam in the episode "The Battle"; Picard pointed out that the ship they were towing would be carried along beside them by its own momentum—its inertia. By the same token, if the *Enterprise* were to attempt to use the tractor beam to ward off the *Stargazer,* the resulting force would push the *Enterprise* backward as effectively as it would push the *Stargazer* forward.

This phenomenon has already dramatically affected the way we work in space at present. Say, for example, that you are an astronaut assigned to tighten a bolt on the Hubble Space Telescope. If you take an electric screwdriver with you to do the job, you are in for a rude awakening after you drift over to the offending bolt. When you switch on the screwdriver as it is pressed against the bolt, you are as likely to start spinning around as the bolt is to turn. This is because the Hubble Telescope is a lot heavier than you are. When the screwdriver applies a force to the bolt, the reaction force you feel may more easily turn you than the bolt, especially if the bolt is still fairly tightly secured to the frame. Of course, if you are lucky enough, like the assassins of Chancellor Gorkon, to have gravity boots that secure you snugly to whatever you are standing on, then you can move about as efficiently as we are used to on Earth.

Likewise, you can see what will happen if the *Enterprise* tries to pull another spacecraft toward it. Unless the *Enterprise* is very much heavier, it will move toward the other object when the tractor beam or grappler turns on, rather than vice versa. In the depths of space, this distinction is a meaningless semantic one. With no reference system nearby, who is to say who is pulling whom? However, if you are on a hapless planet like Moab IV in the path of a renegade star, it makes a great deal of difference whether the *Enterprise* pushes the star aside or the star pushes the *Enterprise* aside!

One trekker I know claims that the way around this problem is already stated indirectly in at least one episode: if the *Enterprise* were to use its impulse engines at the same time that it turned its tractor beam on, it could, by applying an opposing force with its own engines, compensate for any recoil it might feel when it pushed or pulled on something.

This trekker claims that somewhere it is stated that the tractor beam requires the impulse drive to be operational in order to work. I, however, have never noticed any instructions from Kirk or Picard to turn on the impulse engines at the same time the tractor beam is used. And in fact, for a society capable of designing and building inertial dampers, I don't think such a brute force solution would be necessary. Reminded of Geordi LaForge's need for a warp field to attempt to push back the moon at Bre'el IV, I think a careful, if presently unattainable, manipulation of space and time would do the trick equally well. To understand why, we need to engage the inertial dampers and accelerate to the modern world of curved space and time.

2

Einstein Raises

There once was a lady named Bright,
Who traveled much faster than light.
She departed one day, in a relative way,
And returned on the previous night.

—Anonymous

"Time, the final frontier"—or so, perhaps, each Star Trek episode should begin. Forty years ago, in the classic episode "Tomorrow Is Yesterday," the round-trip time travels of the *Enterprise* began. (Actually, at the end of an earlier episode, "The Naked Time," the *Enterprise* is thrown back in time three days—*but* it is only a one-way trip.) The starship is kicked back to twentieth-century Earth as a result of a close encounter with a "black star" (the term "black hole" having not yet permeated the popular culture). Nowadays exotica like wormholes and "quantum singularities" regularly spice up episodes of *Star Trek: Voyager* and *Enterprise* contributed nothing less than a Temporal Cold War. Thanks to Albert Einstein and those who have followed in his footsteps, the very fabric of spacetime is filled with drama.

While every one of us is a time traveler, the cosmic pathos that elevates human history to the level of tragedy arises precisely because we seem doomed to travel in only one direction—into the future. What wouldn't any of us give to travel into the past, relive glories, correct wrongs, meet our heroes, perhaps even avert disasters, or simply revisit youth with the wisdom of age? The possibilities of space travel beckon us every time we gaze up at the stars, yet we seem to be permanent captives in the present. The question that motivates not only dramatic license but a surprising amount of modern theoretical physics research can be simply put: Are we or are we not prisoners on a cosmic temporal freight train that cannot jump the tracks?

The origins of the modern genre we call science fiction are closely tied to the issue of time travel. Mark Twain's early classic *A Connecticut Yankee in King Arthur's Court* is more a work of fiction than science fiction, in spite of the fact that the whole piece revolves around the time-travel adventures of a hapless American in medieval England. (Perhaps Twain did not dwell longer on the scientific aspects of time travel because of the promise he made to Picard aboard the *Enterprise* not to reveal his glimpse of the future once he returned to the nineteenth century by jumping through a temporal rift on Devidia II, in the episode "Time's Arrow.") But H. G. Wells's remarkable work *The Time Machine* completed the transition to the paradigm that Star Trek has followed. Wells was a graduate of the Imperial College of Science and Technology, in London, and scientific language permeates his discussions, as it does the discussions of the *Enterprise* crew.

Surely among the most creative and compelling episodes in the Star Trek series are those involving time travel. I have counted no less than twenty-two episodes in the first two series which deal with this theme, and so do three of the Star Trek movies and a number of the episodes of *Voyager* and *Deep Space Nine* and most recently of *Enterprise,* where time travel played a very central continuing plot role. Perhaps the most fascinating aspect of time travel as far as Star Trek is concerned is that there is no stronger potential for violation of the Prime Directive. The crews of Starfleet are admonished not to interfere with the present normal historical development of any alien society they visit. Yet by traveling back in time it is possible to remove the present altogether. Indeed, it is possible to remove history altogether!

A famous paradox is to be found in both science fiction and physics: What happens if you go back in time and kill your mother before you were born? You must then cease to exist. But if you cease to exist, you could not have gone back and killed your mother. But if you didn't kill your mother, then you have not ceased to exist. Put another way: if you exist, then you cannot exist, while if you don't exist, you must exist.

There are other, less obvious but equally dramatic and perplexing questions that crop up the moment you think about time travel. For example, at the resolution of "Time's Arrow," Picard ingeniously sends a message from the nineteenth to the twenty-fourth century by tapping binary code into Data's severed head, which he knows will be discovered almost five hundred years later and reattached to Data's body. As we watch, he taps the message, and then we cut to LaForge in the twenty-fourth century, as he succeeds in reattaching Data's head. To the viewer these events seem contemporaneous, but they are not; once Picard has tapped the message into Data's head, it lies there for half a millennium. But if I were carefully examining Data's head in the twenty-fourth century and Picard had not yet traveled back in time to change the future, would I see such a message? One might argue that if Picard hasn't traveled back in time yet, there can have been no effect on Data's head. Yet the actions that change Data's programming were performed in the nineteenth century regardless of when Picard traveled back in time to perform them. Thus they have already happened, even if Picard has not yet left! In this way, a cause in the nineteenth century (Picard tapping) can produce an effect in the twenty-fourth century (Data's circuitry change) before the cause in the twenty-fourth century (Picard leaving the ship) produces the effect in the nineteenth century (Picard's arrival in the cave where Data's head is located) which allowed the original cause (Picard tapping) to take place at all.

These confusions are of course nothing compared to the time travel gymnastics associated with the Temporal Cold

War that Captain Archer deals with in *Enterprise,* but as confusing as those multiple timelines are they may still pale in comparison to the Mother of all time paradoxes, which arises in the final episode of *Star Trek: The Next Generation,* when Picard sets off a chain of events that will travel back in time and destroy not just his own ancestry but all life on Earth. Specifically, a "subspace temporal distortion" involving "antitime" threatens to grow backward in time, eventually engulfing the amino acid protoplasm on the nascent Earth before the first proteins, which will be the building blocks of life, can form. This is the ultimate case of an effect producing a cause. The temporal distortion is apparently created in the future. If, in the distant past, the subspace temporal distortion was able to destroy the first life on Earth, then life on Earth could never have evolved to establish a civilization capable of creating the distortion in the future!

The standard resolution of these paradoxes, at least among many physicists, is to argue a priori that such possibilities must not be allowed in a sensible universe, such as the one we presumably live in. However, the problem is that Einstein's equations of general relativity not only do not directly forbid such possibilities, they encourage them.

Within thirty years of the development of the equations of general relativity, an explicit solution in which time travel could occur was developed by the famous mathematician Kurt Gödel, who worked at the Institute for Advanced Study in Princeton along with Einstein. In Star Trek language, this solution allowed the creation of a "temporal causality loop," such as the one the *Enterprise* got caught in after being hit by the *Bozeman*. The dryer terminology of modern physics labels this a "closed timelike curve." In either case, what it implies is that you can travel on a round-trip and return to your starting point in both space *and* time! Gödel's solution involved a universe that, unlike the one we happen to live in, is not expanding but instead is spinning uniformly. In such a universe, it turns out that one could in principle go back in time merely by traveling in a large circle in space. While such a hypothetical universe is dramatically different than the one in which we live, the mere fact that this solution exists at all indicates clearly that time travel is possible within the context of general relativity.

There is a maxim about the universe that I always tell my students: That which is not explicitly forbidden is guaranteed to occur. Or, as Data said in the episode "Parallels," referring to the laws of quantum mechanics, "All things which can occur, do occur." This is the spirit with which I think one should approach the physics of Star Trek. We must consider the distinction not between what is practical and what is not, but between what is possible and what is not.

This fact was not, of course, lost on Einstein himself, who wrote, "Kurt Gödel's [time machine solution raises] the problem [that] disturbed me already at the time of the building up of the general theory of relativity, without my having succeeded in clarifying it. . . . It will be interesting to weigh whether these [solutions] are not to be excluded on physical grounds."[1]

The challenge to physicists ever since has been to determine what if any "physical grounds" exist that would rule out the possibility of time travel, which the form of the equations of general relativity appears to foreshadow. To discuss such things will require us to travel beyond the classical world of general relativity to a murky domain where quantum mechanics must affect even the nature of space and time. On the way, we, like the *Enterprise,* will encounter black holes and wormholes. But first we ourselves must travel back in time to the latter half of the nineteenth century.

The marriage of space and time that heralded the modern era began with the marriage, in 1864, of electricity and magnetism. This remarkable intellectual achievement, based on the cumulative efforts of great physicists such as André-Marie Ampère, Charles-Augustin de Coulomb, and Michael Faraday, was capped by the brilliant British physicist James Clerk Maxwell. He discovered that the laws of electricity and magnetism not only displayed an intimate relationship with one another but together implied the existence of "electromagnetic waves," which should travel throughout space at a speed that one could calculate based on the known properties of electricity and magnetism. The speed turned out to be identical to the speed of light, which had previously been measured.

Now, since the time of Newton there had been a debate about whether light was a wave—that is, a traveling disturbance in some background medium—or a particle, which travels regardless of the presence of a background medium.

The observation of Maxwell that electromagnetic waves must exist and that their speed was identical to that of light ended the debate: light was an electromagnetic wave.

Any wave is just a traveling disturbance. Well, if light is an electromagnetic disturbance, then what is the medium that is being disturbed as the wave travels? This became the hot topic for investigation at the end of the nineteenth century. The proposed medium had had a name since Aristotle. It was called the aether, and had thus far escaped any attempts at direct detection. In 1887, however, Albert A. Michelson and Edward Morley, working at the institutions that later merged in 1967 to form my present home, Case Western Reserve University, performed an experiment guaranteed to detect not the aether but the aether's effects: Since the aether was presumed to fill all of space, the Earth was presumed to be in motion through it. Light traveling in different directions with respect to the Earth's motion through the aether ought therefore to show variations in speed. This experiment has since become recognized as one of the most significant of the last century, even though Michelson and Morley never observed the effect they were searching for. In fact, it is precisely because they failed to observe the effect of the Earth's motion through the aether that we remember their names today. (A. A. Michelson actually went on to become the first American Nobel laureate in physics for his experimental investigations into the speed of light, and I feel privileged to hold a position today which he held more than a hundred years ago. Edward Morley continued as a renowned chemist and determined the atomic weight of helium, among other things.)

The nondiscovery of the aether did send minor ripples of shock throughout the physics community, but, like many watershed discoveries, its implications were fully appreciated only by a few individuals who had already begun to recognize several paradoxes associated with the theory of electromagnetism. Around this time, a young high school student who had been eight years old at the time of the Michelson-Morley experiment independently began to try to confront these paradoxes directly. By the time he was twenty-six, in the year 1905, Albert Einstein had solved the problem. But as also often occurs whenever great leaps are made in physics, Einstein's results created more questions than they answered.

Einstein's solution, forming the heart of his special theory of relativity, was based on a simple but apparently impossible fact: the only way in which Maxwell's theory of electromagnetism could be self-consistent would be if the observed speed of light was independent of the observer's speed relative to the light. The problem, however, is that this completely defies common sense. If a probe is released from the *Enterprise* when the latter is traveling at impulse speed, an observer on a planet below will see the probe whiz past at a much higher speed than would a crew member looking out an observation window on the *Enterprise*. However,

Einstein recognized that Maxwell's theory would be self-consistent only if light waves behaved differently—that is, if their speed as measured by both observers remained identical, independent of the relative motion of the observers. Thus, if I shoot a phaser beam out the front of the *Enterprise,* and it travels away from the ship at the speed of light toward the bridge of a Romulan Warbird, which itself is approaching the *Enterprise* at an impulse speed of 3/4 the speed of light, those on the enemy bridge will observe the beam to be heading toward them just at the speed of light and not at 1 3/4 times the speed of light. This sort of thing has confused some trekkers, who imagine that if the *Enterprise* is moving at near light speed and another ship is moving in the opposite direction at near light speed, the light from the *Enterprise* will never catch up with the other ship (and therefore the *Enterprise* will not be visible to it). Instead, those on the other ship will see the light from the *Enterprise* approaching at the speed of light.

This realization alone was not what made Einstein's a household name. More important was the fact that he was willing to explore the implications of this realization, all of which on the surface seem absurd. In our normal experience, it is time and space that are absolute, while speed is a relative thing: how fast something is perceived to be moving depends upon how fast you yourself are moving. But as one approaches light speed, it is speed that becomes an absolute quantity, and therefore *space and time must become relative!*

This comes about because speed is literally defined as distance traveled during some specific time. Thus, the only way observers in relative motion can measure a single light ray to traverse the same distance—say, 300 million meters—relative to each of them in, say, one second is if each of their "seconds" is different or each of their "meters" is different! It turns out that in special relativity, the "worst of both worlds" happens—that is, seconds and meters both become relative quantities.

From the simple fact that the speed of light is measured to be the same for all observers, regardless of their relative motion, Einstein obtained the four following consequences for space, time, and matter:

(a) Events that occur for one observer *at the same time in two different places* need not be simultaneous to another observer moving with respect to the first. *Each person's "now" is unique to themselves. "Before" and "after" are relative for distant events.*
(b) All clocks on starships that are moving relative to me will appear to me to be ticking more slowly than my clock. *Time is measured to slow down for objects in motion.*
(c) All yardsticks on starships that are moving relative to me will appear shorter than they would if they were standing still in my frame. *Objects, including starships, are measured to contract if they are moving.*
(d) All massive objects get heavier the faster they travel. As they approach the speed of light, they become

infinitely heavy. *Thus, only massless objects, like light, can actually travel at the speed of light.*

This is not the place to review all of the wonderful apparent paradoxes that relativity introduces into the world. Suffice it to say that, like it or not, consequences (a) through (d) are true—that is, they have been tested. Atomic clocks have been carried aloft in high-speed aircraft and have been observed to be behind their terrestrial counterparts upon their return. In high-energy physics laboratories around the world, the consequences of the special theory of relativity are the daily bread and butter of experiment. Unstable elementary particles are made to move near the speed of light, and their lifetimes are measured to increase by huge factors. When electrons, which at rest are 2000 times less massive than protons, are accelerated to near light speed, they are measured to carry a momentum equivalent to that of their heavier cousins. Indeed, an electron accelerated to .99999999999999999999999999999999999999999999999999 9999999999999 times the speed of light would hit you with the same impact as a Mack truck traveling at normal speed.

Of course, the reason all these implications of the relativity of space and time are so hard for us to accept at face value is that we happen to live and move at speeds far smaller than the speed of light. Each of the above effects becomes noticeable only when one is moving at "relativistic" speeds. For example, even at half the speed of light, clocks would slow and yardsticks would shrink by only about 15 percent. On NASA's space shuttle, which moves at about 5 miles per second around the Earth, clocks tick less than one ten-millionth of a percent slower than their counterparts on Earth.

However, in the high-speed world of the *Enterprise* or any other starship, relativity would have to be confronted on a daily basis. Indeed, in managing a Federation, one can imagine the difficulties of synchronizing clocks across a large segment of the galaxy when a great many of these clocks are moving at close to light speed. As a result, Starfleet apparently has a rule that normal impulse operations for starships are to be limited to a velocity of 0.25 *c*—that is, 1/4 light speed, or a mere 75,000 km/sec.[2]

Even with such a rule, clocks on ships traveling at this speed will slow by slightly over 3 percent compared with clocks at Starfleet Command. This means that in a month of travel, clocks will have slowed by almost one day. If the *Enterprise* were to return to Starfleet Command after such a trip, it would be Friday on the ship but Saturday back home. I suppose the inconvenience would not be any worse than resetting your clocks after crossing the international date line when traveling to Asia, except in this case the crew would *actually be* one day younger after the round-trip, whereas on a round-trip to the Orient you gain one day going in one direction and lose one going in the other.

You can now see how important warp drive is to the *Enterprise*. Not only is it designed to avoid the ultimate speed limit—the speed of light—and so allow practical travel across the galaxy, but it is also designed to avoid the problems of time dilation, which result when the ship is traveling close to light speed.

I cannot overemphasize how significant these facts are. The fact that clocks slow down as one approaches the speed of light has been taken by science fiction writers (and indeed by all those who have dreamed of traveling to the stars) as opening the possibility that one might cross the vast distances between the stars in a human lifetime—at least a human lifetime for those aboard the spaceship. At close to the speed of light, a journey to, say, the center of our galaxy would take more than 25,000 years of Earth time. For those aboard the spaceship, if it were moving sufficiently close to light speed, the trip might take less than 10 years—a long time, but not impossibly so. Nevertheless, while this might make individual voyages of discovery possible, it would make the task of running a Federation of civilizations scattered throughout the galaxy impossible. As the writers of Star Trek have correctly surmised, the fact that a 10-year journey for the *Enterprise* would correspond to a 25,000-year period for Starfleet Command would wreak havoc on any command operation that hoped to organize and control the movements of many such craft. Thus it is absolutely essential that (a) light speed be avoided, in order not to put the Federation out of synchronization, *and* (b) faster-than-light speed be realized, in order to move practically about the galaxy.

The kicker is that, in the context of special relativity alone, the latter possibility *cannot be realized.* Physics becomes full of impossibilities if super light speed is allowed. Not least among the problems is that because objects get more massive as they approach the speed of light, it takes progressively more and more energy to accelerate them by a smaller and smaller amount. As in the myth of the Greek hero Sisyphus, who was condemned to push a boulder uphill for all eternity only to be continually thwarted near the very top, all the energy in the universe would not be sufficient to allow us to push even a speck of dust, much less a starship, past this ultimate speed limit.

By the same token, not just light but all massless radiation *must* travel at the speed of light. This means that the many types of beings of "pure energy" encountered by the *Enterprise,* and later by the *Voyager,* would have difficulty existing as shown. In the first place, they wouldn't be able to sit still. Light cannot be slowed down, let alone stopped in empty space. In the second place, any form of intelligent-energy being (such as the "photonic" energy beings in the *Voyager* series; the energy beings in the Beta Renna cloud, in *The Next Generation;* the Zetarians, in the original series; and the Dal'Rok, in *Deep Space Nine*), which is constrained to travel at the speed of light, would have clocks that are infinitely slowed compared to our own. The entire history of the universe would pass by in a single instant. If energy beings could experience anything, they would experience everything at once! Needless to say, before they could actually interact with corporeal beings the corporeal beings would be long dead.

Speaking of time, I think it is time to introduce the Picard Maneuver. Jean-Luc became famous for introducing this tactic while stationed aboard the *Stargazer.* Even though it involves warp travel, or super light speed, which I have argued is impossible in the context of special relativity alone, it does so for just an instant and it fits in nicely with the discussions here. In the Picard Maneuver, in order to confuse an attacking enemy vessel, one's own ship is accelerated to warp speed for an instant. It then appears to be in two places at once. This is because, traveling faster than the speed of light for a moment, it *overtakes* the light rays that left it the instant before the warp drive was initiated. While this is a brilliant strategy—and it appears to be completely consistent as far as it goes (that is, ignoring the issue of whether it is possible to achieve warp speed)—I think you can see that it opens a veritable Pandora's can of worms. In the first place, it begs a question that has been raised by many trekkers over the years: How can the *Enterprise* bridge crew "see" objects approaching them at warp speed? Just as surely as the *Stargazer* overtook its own image, so too will all objects traveling at warp speed; one shouldn't be able to see the moving image of a warp-speed object until long after it has arrived. One can only assume that when Kirk, Picard, Janeway, or Archer orders up an image on the viewscreen, the result is an image assembled by some sort of long-range "subspace" (that is, super-light-speed communication) sensors. Even ignoring this apparent oversight, the Star Trek universe would be an interesting and a barely navigable one, full of ghost images of objects that long ago arrived where they were going at warp speed.

⊂⊃

Moving back to the sub-light-speed world: We are not through with Einstein yet. His famous relation between mass and energy, $E=mc^2$, which is a consequence of special relativity, presents a further challenge to space travel at impulse speeds. As I have described it in chapter 1, a rocket is a device that propels material backward in order to move forward. As you might imagine, the faster the material is propelled backward, the larger will be the forward impulse the rocket will receive. Material cannot be propelled backward any faster than the speed of light. Even propelling it at light speed is not so easy: the only way to get propellant moving backward at light speed is to make the fuel out of matter and antimatter, which (as I describe in a later chapter) can completely annihilate to produce pure radiation moving at the speed of light.

However, while the warp drive aboard the *Enterprise* uses such fuel, the impulse drive does not. It is powered instead by nuclear fusion—the same nuclear reaction that powers the Sun by turning hydrogen into helium. In fusion reactions, about 1 percent of the available mass is converted into energy. With this much available energy, the helium atoms that are produced can come streaming out the back of the rocket at about an eighth of the speed of light. Using this exhaust velocity for the propellant, we then can calculate the amount of fuel the *Enterprise* needs in order to accelerate to, say, half the speed of light. The calculation is not difficult, but I will just give the answer here. It may surprise you. Each time the *Enterprise* accelerates to half the speed of light, it must burn *81 TIMES ITS ENTIRE MASS* in hydrogen fuel. Given that a Galaxy Class starship such as Picard's *Enterprise-D* would weigh in excess of 4 million metric tons,[3] this means that over 300 million metric tons of fuel would need to be used each time the impulse drive is used to accelerate the ship to half light speed! If one used a matter-antimatter propulsion system for the impulse drive, things would be a little better. In this case, one would have to burn merely *twice* the entire mass of the *Enterprise* in fuel for each such acceleration.

It gets worse. The calculation I described above is correct for a single acceleration. To bring the ship to a stop at its destination would require the same factor of 81 times its mass in fuel. Since one must carry the fuel needed to stop aboard the ship during the period in which one is trying to speed up to half light speed, this means that just to go somewhere at half light speed and stop again would require fuel in the amount of 81 x (ship's mass plus fuel for stopping carried along on outbound voyage) = 81 x 82 = 6642 *TIMES THE ENTIRE SHIP'S MASS!* Moreover, say that one wanted to achieve the acceleration to half the speed of light in a few hours (we will assume, of course, that the inertial dampers are doing their job of shielding the crew and ship from the tremendous G-forces that would otherwise ensue). The power radiated as propellant by the engines would then be about 10^{22} watts—or about a billion times the total average power presently produced and used by all human activities on Earth!

Now, you may suggest (as a bright colleague of mine did when I originally presented him with this argument) that there is a subtle loophole. The argument hinges on the requirement that you carry your fuel along with the rocket. What if, however, you harvest your fuel as you go along? After all, hydrogen is the most abundant element in the universe. Can you not sweep it up as you move through the galaxy? Well, the average density of matter in our galaxy is about one hydrogen atom per cubic centimeter. To sweep up just one gram of hydrogen per second, even moving at a good fraction of the speed of light, would require you to deploy collection panels with a diameter of over 25 miles. And even turning all this matter into energy for propulsion would provide only about a hundred-millionth of the needed propulsion power!

To paraphrase the words of the Nobel Prize–winning physicist Edward Purcell, whose arguments I have adapted and extended here: If this sounds preposterous to you, you are right. Its preposterousness follows from the elementary laws of classical mechanics and special relativity. The arguments presented here are as inescapable as the fact that a ball will fall when you drop it at the Earth's surface. Rocket-propelled space travel through the galaxy at near light speed *is not physically practical,* now or ever!

So, do I end the book here? Do we send back our Star Trek com badges and ask for a refund? Well, we are still not done with Einstein. His final, perhaps greatest discovery holds out a glimmer of hope after all.

Fast rewind back to 1908: Einstein's discovery of the relativity of space and time heralds one of those "Aha!" experiences that every now and then forever change our picture of the universe. It was in the fall of 1908 that the mathematical physicist Hermann Minkowski wrote these famous words: "Henceforth, space by itself, and time by itself, are doomed to fade away into mere shadows, and only a kind of union of the two will preserve an independent reality."

What Minkowski realized is that even though space and time are relative for observers in relative motion—your clock can tick slower than mine, and my distances can be different from yours—if space and time are instead merged as part of a four-dimensional whole (three dimensions of space and one of time), an "absolute" objective reality suddenly reappears.

The leap of insight Minkowski had can be explained by recourse to a world in which everyone has monocular vision and thus no direct depth perception. If you were to close one eye, so that your depth perception was reduced, and I were to hold a ruler up for you to see, and I then told someone else, who was observing from a different angle, to close one eye too, the ruler I was holding up would appear to the other observer to be a different length than it would appear to be to you—as the following bird's-eye view shows.

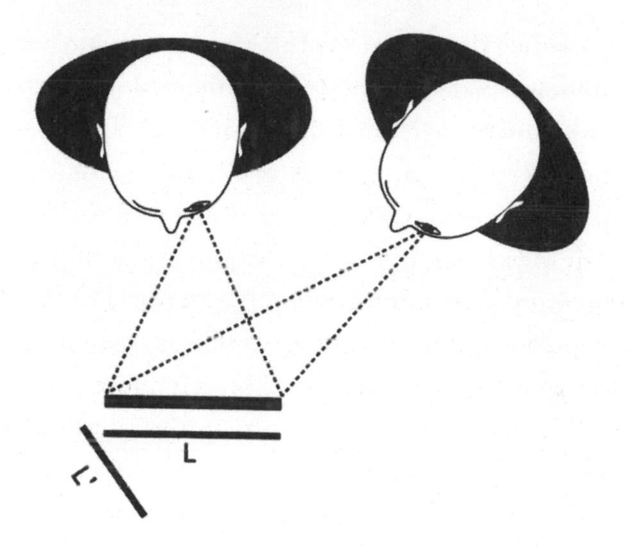

Each observer in the example above, without the direct ability to discern depth, will label "length" (L or L') to be the two-dimensional projection onto his or her plane of vision of the actual three-dimensional length of the ruler. Now, because we know that space has three dimensions, we are not fooled by this trick. We know that viewing something from a different angle does not change its real length, even if it changes its apparent length. Minkowski showed that the same idea can explain the various paradoxes of relativity, if we now instead suppose that our perception of space is merely a three-dimensional slice of what is actually a four-dimensional manifold in which space and time are joined. Two different observers in relative motion perceive *different* three-dimensional slices of the underlying four-dimensional space in much the same way that the two rotated observers pictured on the previous page view *different* two-dimensional slices of a three-dimensional space.

Minkowski imagined that the spatial distance measured by two observers in relative motion is a projection of an underlying *four-dimensional spacetime distance* onto the three-dimensional space that they can sense; and, similarly, that the temporal "distance" between two events is a projection of the four-dimensional spacetime distance onto their own timeline. Just as rotating something in three dimensions can mix up width and depth, so relative motion in four-dimensional space can mix up different observers' notions of "space" and "time." Finally, just as the length of an object does not change when we rotate it in space, the four-dimensional spacetime distance between two events is absolute—independent of how different observers in relative motion assign "spatial" and "temporal" distances.

So the crazy invariance of the speed of light for all observers provided a key clue to unravel the true nature of the four-dimensional universe of spacetime in which we actually live. *Light displays the hidden connection between space and time.* Indeed, the speed of light *defines* the connection.

It is here that Einstein returned to save the day for Star Trek. Once Minkowski had shown that spacetime in special relativity was like a four-dimensional sheet of paper, Einstein spent the better part of the next decade flexing his mathematical muscles until he was able to bend that sheet, which in turn allows us to bend the rules of the game. As you may have guessed, light was again the key.

3

Hawking Shows His Hand

"How little do you mortals understand time. Must you be so linear, Jean-Luc?"

—Q to Picard, in "All Good Things. . . . "

The planet Vulcan, home to Spock, actually has a venerable history in twentieth-century physics. A great puzzle in astrophysics in the early part of this century was the fact that the perihelion of Mercury—the point of its closest approach to the Sun—was precessing around the Sun each Mercurian year by a very small amount in a way that was not consistent with Newtonian gravity. It was suggested that a new planet existed inside Mercury's orbit that could perturb it in such a way as to fix the problem. (In fact, the same solution to an anomaly in the orbit of Uranus had earlier led to the discovery of the planet Neptune.) The name given to the hypothetical planet was Vulcan.

Alas, the mystery planet Vulcan is not there. Instead, Einstein proposed that the flat space of Newton and Minkowski had to be given up for the curved spacetime of general relativity. In this curved space, Mercury's orbit would deviate slightly from that predicted by Newton, explaining the observed discrepancy. While this removed the need for the planet Vulcan, it introduced possibilities that are much more exciting. Along with curved space come black holes, wormholes, and perhaps even warp speeds and time travel.

Indeed, long before the Star Trek writers conjured up warp fields, Einstein warped spacetime, and, like the Star Trek writers, he was armed with nothing other than his imagination. Instead of imagining twenty-second-century starship technology, however, Einstein imagined an elevator. He was undoubtedly a great physicist, but he probably never would have sold a screenplay.

Nonetheless, his arguments remain intact when translated aboard the *Enterprise.* Because light is the thread that weaves together space and time, the trajectories of light rays give us a map of spacetime just as surely as warp and weft threads elucidate the patterns of a tapestry. Light generally travels in straight lines. But what if a Romulan commander aboard a nearby Warbird shoots a phaser beam at Picard as he sits on the bridge of his captain's yacht *Calypso,* having just engaged the impulse drive (we will assume the inertial dampers are turned off for this example)? Picard would accelerate forward, narrowly missing the brunt of the phaser blast. When viewed in Picard's frame of reference, things would look like the figure at the top of the following page.

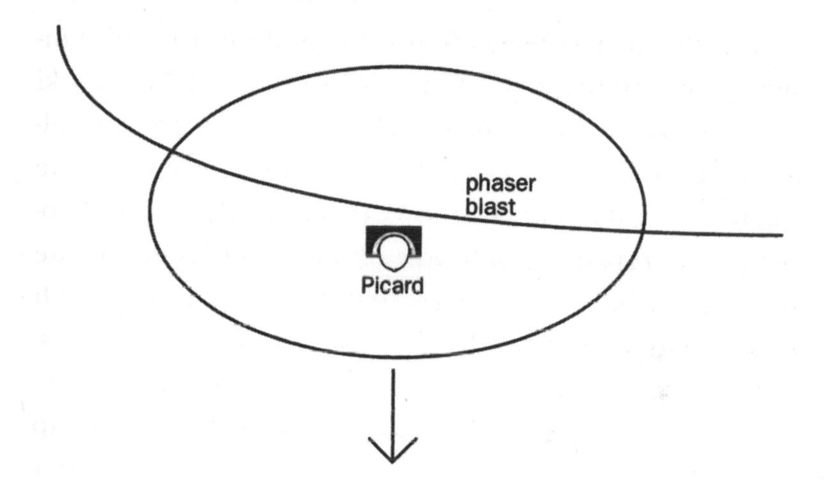

So, for Picard, the trajectory of the phaser ray would be curved. What else would Picard notice? Well, recalling the argument in the first chapter, as long as the inertial dampers are turned off, he would be thrust back in his seat. In fact, I also noted there that if Picard was being accelerated forward at the same rate as gravity causes things to accelerate downward at the Earth's surface, he would feel exactly the same force pushing him back against his seat that he would feel pushing him down if he were standing on Earth. In fact, Einstein argued that Picard (or his equivalent in a rising elevator) would never be able to perform any experiment that could tell the difference between the reaction force due to his acceleration and the pull of gravity from some nearby heavy object outside the ship. Because of this, Einstein boldly went where no physicist had gone before and reasoned that whatever phenomena an accelerating observer experienced would be identical to the phenomena an observer in a gravitational field experienced.

Our example implies the following: Since Picard observes the phaser ray bending when he is accelerating away from it, the ray must also bend in a gravitational field. But if light rays map out spacetime, then *spacetime* must bend in a gravitational field. Finally, since matter produces a gravitational field, then *matter must bend spacetime!*

Now, you may argue that since light has energy, and mass and energy are related by Einstein's famous equation, then the fact that light bends in a gravitational field is no big surprise—and certainly doesn't seem to imply that we have to believe that spacetime itself need be curved. After all, the paths that matter follows bend too (try throwing a ball in the air). Galileo could have shown, had he known about such objects, that the trajectories of baseballs and Pathfinder missiles bend, but he never would have mentioned curved space.

Well, it turns out that you can calculate how much a light ray should bend if light behaved the same way a baseball does, and then you can go ahead and measure this bending, as Sir Arthur Stanley Eddington did in 1919 when he led an expedition to observe the apparent position of stars on the sky very near the Sun during a solar eclipse. Remarkably, you would find, as Eddington did, that light bends exactly *twice* as much as Galileo might have predicted if it behaved like a baseball in flat space. As you may have guessed, this factor of 2 is just what Einstein predicted if spacetime was curved in the vicinity of the Sun and light (or the planet Mercury, for that matter) was locally traveling in a straight line in this curved space! Suddenly, Einstein's was a household name.

Curved space opens up a whole universe of possibilities, if you will excuse the pun. Suddenly we, and the *Enterprise,* are freed from the shackles of the kind of linear thinking imposed on us in the context of special relativity, which Q, for one, seemed to so abhor. One can do many things on a curved manifold which are impossible on a flat one. For example, it is possible to keep traveling in the same direction and yet return to where you began—people who travel around the world do it all the time.

The central premise of Einstein's general relativity is simple to state in words: the curvature of spacetime is directly determined by the distribution of matter and energy contained within it. Einstein's equations, in fact, provide simply the strict mathematical relation between curvature on the one hand and matter and energy on the other:

$$\underset{\{\text{CURVATURE}\}}{\text{Left-hand side}} = \underset{\{\text{MATTER AND ENERGY}\}}{\text{Right-hand side}}$$

What makes the theory so devilishly difficult to work with is this simple feedback loop: the curvature of spacetime is determined by the distribution of matter and energy in the universe, but this distribution is in turn governed by the curvature of space. It is like the chicken and the egg. Which was there first? Matter acts as the source of curvature, which in turn determines how matter evolves, which in turn alters the curvature, and so on.

Indeed, this may be perhaps the most important single aspect of general relativity as far as Star Trek is concerned. The complexity of the theory means that we still have not yet fully understood all its consequences; therefore we cannot rule out various exotic possibilities. It is these exotic possibilities that are the grist of Star Trek's mill. In fact, we shall see that all these possibilities rely on one great unknown that permeates everything, from wormholes and black holes to time machines.

The first implication of the fact that spacetime need not be flat that will be important to the adventures of the *Enterprise* is that time itself becomes an even more dynamic quantity than it was in special relativity. Time can flow at different rates for different observers even if they are not moving relative to each other. Think of the ticks of a clock as the ticks on a ruler made of rubber. If I were to stretch or bend the ruler, the spacing between the ticks would differ from point to point. If this spacing represents the ticks of a clock, then clocks located in different places can tick at different rates. In general relativity, the only way to "bend" the ruler is for a gravitational field to be present, which in turn requires the presence of matter.

To translate this into more pragmatic terms: if I put a heavy iron ball near a clock, it should change the rate at which the clock ticks. Or more practical still, if I sleep with my alarm clock tucked next to my body's rest mass, I will be awakened a little later than I would otherwise, at least as far as the rest of the world is concerned.

A famous experiment done in the physics laboratories at Harvard University in 1960 first demonstrated that time can depend on where you are. Robert Pound and George Rebka showed that the frequency of gamma radiation measured at its source, in the basement of the building, differed from the frequency of the radiation when it was received 74 feet higher, on the building's roof (with the detectors having been carefully calibrated so that any observed difference would not be detector-related). The shift was an incredibly small amount—about 1 part in a million billion. If each cycle of the gamma-ray wave is like the tick of an atomic clock, this experiment implies that a clock in the basement will appear to be running more slowly than an equivalent atomic clock on the roof. Time slows on the lower floor because this is closer to the Earth than the roof is, so the gravitational field, and hence the spacetime curvature, is larger there. As small as this effect was, it was precisely the value predicted by general relativity, assuming that spacetime is curved near the Earth.

In fact, while this effect is small, it is nevertheless of vital importance in modern daily life. The GPS devices in cars today receive signals from satellites located in high orbits around the Earth. By comparing the arrival times of the pulses from the different satellites, the device can pinpoint precisely its location on the Earth's surface. However, the different satellites are in different orbits, and their clocks are ticking at different rates because of this. While the effects are miniscule, they are significant on the scale relevant for GPS positional accuracy. Thus the clocks must be corrected for the effects of gravity or the satellite tracking systems would go out of alignment in a matter of seconds.

The second implication of curved space is perhaps even more exciting as far as space travel is concerned. If space is curved, then a straight line need not be the shortest distance between two points. Here's an example. Consider a circle on a piece of paper. Normally, the shortest distance between two points A and B located on opposite sides of the circle is given by the line connecting them through the center of the circle:

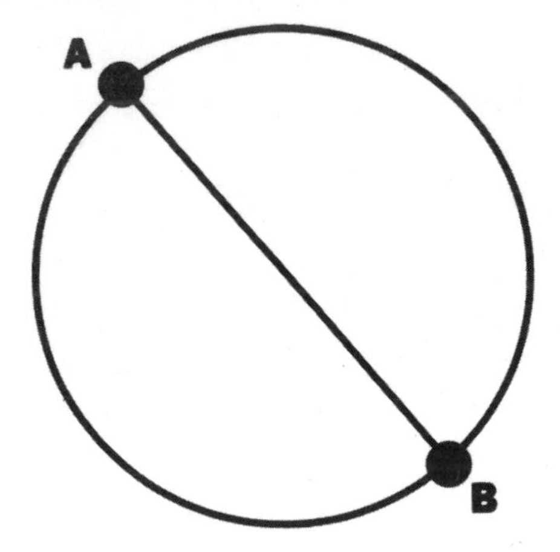

If, instead, one were to travel around the circle to get from A to B, the journey would be about 1 1/2 times as long. However, let me draw this circle on a rubber sheet and distort the central region:

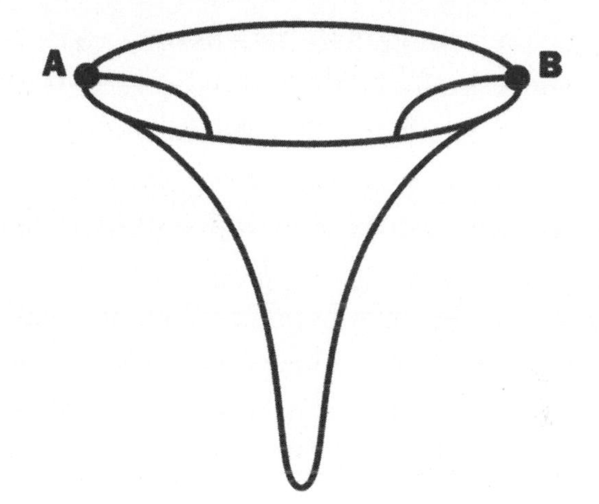

Now, when viewed in our three-dimensional perspective, it is clear that the journey from A to B taken through the center of the region will be much longer than that taken by going around the circle. Note that if we took a snapshot of this from above, so we would have only a two-dimensional perspective, the line from A to B through the center would look like a straight line. More relevant perhaps, if a tiny bug (or two-dimensional beings, of the type encountered by the *Enterprise*) were to follow the trajectory from A to B through the center by crawling along the surface of the sheet, this trajectory would appear to be straight. The bug would be amazed to find that the straight line through the center between A and B was no longer the shortest distance between these two points. If the bug were intelligent, it would be forced to the conclusion that the two-dimensional space it lived in was curved. Only by viewing the embedding of this sheet in the underlying three-dimensional space can we observe the curvature directly.

Now, remember that we live within a four-dimensional spacetime that can be curved, and we can no more perceive the curvature of this space directly than the bug crawling on the surface of the sheet can detect the curvature of the sheet. I think you know where I am heading: If, in curved space, the shortest distance between two points need not be a straight line, then it might be possible to traverse what appears *along the line of sight* to be a huge distance, by finding instead a shorter route through curved spacetime.

These properties I have described are the stuff that Star Trek dreams are made of. Of course, the question is: How many of these dreams may one day come true?

Wormholes: Fact and Fancy. The Bajoran wormhole in *Deep Space Nine* is perhaps the most famous wormhole in Star Trek, although there have been plenty of others, including the dangerous wormhole that Scotty could create by imbalancing the matter-antimatter mix in the *Enterprise*'s warp drive; the unstable Barzan wormhole, through which a Ferengi ship was lost in the *Next Generation* episode "The Price"; and the temporal wormhole that the *Voyager* encountered in its effort to get back home from the far edge of the galaxy.

The idea that gives rise to wormholes is exactly the one I just described. If spacetime is curved, then perhaps there are different ways of connecting two points so that the distance between them is much shorter than that which would be measured by traveling in a "straight line" through curved space. Because curved-space phenomena in four dimensions are impossible to visualize, we once again resort to a two-dimensional rubber sheet, whose curvature we can observe by embedding it in three-dimensional space.

If the sheet is curved on large scales, one might imagine that it looks something like this:

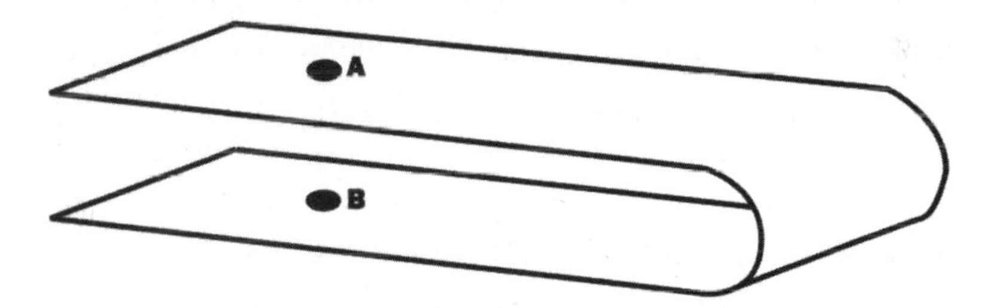

Clearly, if we were to poke a pencil down at A and stretch the sheet until we touched B, and then sewed together the two parts of the sheet, like so:

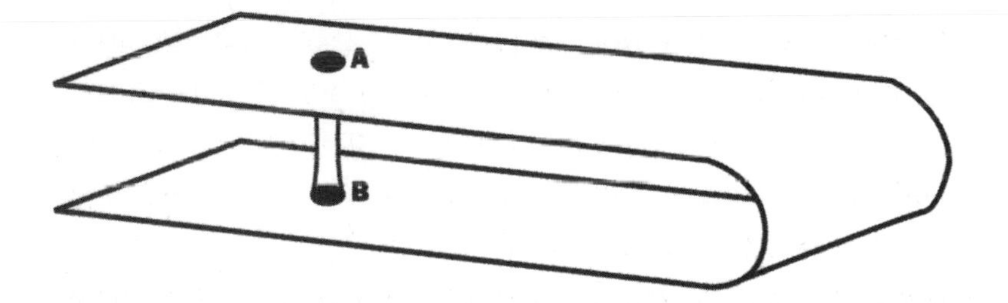

we would create a path from A to B that was far shorter than the path leading around the sheet from one point to another. Notice also that the sheet appears flat near A and also near B. The curvature that brings these two points close enough together to warrant joining them by a tunnel is due to the global bending of the sheet over large distances. A little bug (even an intelligent one) at A, confined to crawl on the sheet, would have no idea that B was as "close" as it was, even if it could do some local experiments around A to check for a curvature of the sheet.

As you have no doubt surmised, the tunnel connecting A and B in this figure is a two-dimensional analogue of a three-dimensional wormhole, which could, in principle, connect distant regions of spacetime. As exciting as this possibility is, there are several deceptive aspects of the picture which I want to bring to your attention. In the first place, even though the rubber sheet is shown embedded in a three-dimensional space in order for us to "see" the curvature of the sheet, the curved sheet can exist without the three-dimensional space around it needing to exist. Thus, while a wormhole could exist joining A and B, there is no sense in which A and B are "close" *without* the wormhole being present. It is not as if one is free to leave the rubber sheet and move from A to B through the three-dimensional space in which the sheet is embedded. If the three-dimensional space is not there, the rubber sheet is all there is to the universe.

Thus, imagine that you were part of an infinitely advanced civilization (but not as advanced as the omnipotent Q beings, who seem to transcend the laws of physics) that had the power to build wormholes in space. Your wormhole building device would effectively be like the pencil in the example I just gave. If you had the power to produce huge local curvatures in space, you would have to poke around blindly in the hope that somehow you could connect two regions of space that, until the instant a wormhole was established, would remain very distant from each other. In no way whatsoever would these two regions be close together until the wormhole produced a bridge. The bridge-building process *itself* is what changes the global nature of spacetime.

Because of this, making a wormhole is not to be taken lightly. When Premier Bhavani of Barzan visited the *Enterprise* to auction off the rights to the Barzan wormhole, she exclaimed, "Before you is the first and only stable wormhole known to exist!" Alas, it wasn't stable; indeed, the only wormholes whose mathematical existence has been consistently established in the context of general relativity are transitory. Such wormholes are created as two microscopic "singularities"—regions of spacetime where the curvature becomes infinitely sharp—find each other and momentarily join. However, in a time shorter than the time it would take a space traveler to pass through such a wormhole, it closes up, leaving once again two disconnected singularities. The unfortunate explorer would be crushed to bits in one singularity or the other before being able to complete the voyage through the wormhole.

The problem of how to keep the mouth of a wormhole open has been hideously difficult to resolve in mathematical detail, but is quite easily stated in physical terms: Gravity sucks! Any kind of normal matter or energy will tend to collapse under its own gravitational attraction unless something else stops it. Similarly, the mouth of a wormhole will pinch off in nothing flat under normal circumstances.

So, the trick is to get rid of the normal circumstances. In recent years, the Caltech physicist Kip Thorne, among others, has argued that the only way to keep wormholes open is to thread them with "exotic material." By this is meant material that will be measured, at least by certain observers, to have "negative" energy. As you might expect (although naive expectations are notoriously suspect in general relativity), such material would tend to "blow," not "suck," as far as gravity is concerned.

Not even a diehard trekker might be willing to suspend disbelief long enough to accept the idea of matter with "negative energy"; however, as noted, in curved space one's normal expectations are often suspect. When you compound this with the exotica forced upon us by the laws of quantum mechanics, which govern the behavior of matter on small scales, quite literally almost all bets are off.

Black Holes and Dr. Hawking. Enter Stephen Hawking. He first became well known among physicists working on general relativity for his part in proving general theorems related to singularities in spacetime, and then, in the 1970s, for his remarkable theoretical discoveries about the behavior of black holes. These objects are formed from material that has collapsed so utterly that the local gravitational field at their surface prevents even light from escaping.

Incidentally, the term "black hole," which has so captivated the popular imagination, was coined by the theoretical physicist John Archibald Wheeler of Princeton University in the late fall of 1967. The date here is very interesting, because, as far as I can determine, the first Star Trek episode to refer to a black hole, which it called a "black star," was aired in 1967 before Wheeler ever used the term in public. When I watched this episode early in the preparation of this book, I found it amusing that the Star Trek writers had gotten the name wrong. Now I realize that they very nearly invented it!

Black holes are remarkable objects for a variety of reasons. First, all black holes eventually hide a spacetime singularity at their center, and anything that falls into the black hole must inevitably encounter it. At such a singularity—an infinitely curved "cusp" in spacetime—the laws of physics as we know them break down. The curvature near the singularity is so large over such a small region that the effects of gravity are governed by the laws of quantum mechanics. Yet no one has yet been able to write down a theory that consistently accommodates both general relativity (that is, gravity) and quantum mechanics. Star Trek writers correctly recognized this tension between quantum mechanics and gravity, as they usually refer to all spacetime singularities as "quantum singularities." One thing is certain, however: by the time the gravitational field at the center of a black hole reaches a strength large enough for our present picture of physics to break down, any ordinary physical object will be torn apart beyond recognition. Nothing could survive intact.

You may notice that I referred to a black hole as "hiding" a singularity at its center. The reason is that at the outskirts of a black hole is a mathematically defined surface we call the "event horizon," which shields our view of what happens to objects that fall into the hole. Inside the event horizon, everything must eventually hit the ominous singularity. Outside the event horizon, objects can escape. While an observer unlucky enough to fall into a black hole will notice nothing special at all as he or she (soon to be "it") crosses the event horizon, an observer watching the process from far away sees something very different. Time slows down for the observer freely falling in the vicinity of the event horizon, relative to an observer located far away. As a result, the falling observer appears from the outside to slow down as he or she nears the event horizon. The closer the falling observer gets to the event horizon, the slower is his or her clock relative to the outside observer's. While it may take the falling observer a few moments (local time) to cross the event horizon—where, I repeat, nothing special happens and nothing special sits—it will take an eternity as observed by someone on the outside. The infalling object appears to become frozen in time.

Moreover, the light emitted by any infalling object gets harder and harder to see from the outside. As an object approaches the event horizon, the object gets dimmer and dimmer (because the observable radiation from it gets shifted to frequencies below the visible). Finally, even if you could see, from the outside, the object's transit of the event horizon (which you cannot witness in any finite amount of time), the object would disappear completely once it passed the horizon, because any light it emitted would be trapped inside, along with the object. Whatever falls inside the event horizon is lost forever to the outside world. It appears that this lack of communication is a one-way street: an observer on the outside can send signals *into* the black hole, but no signal can ever be returned.

For these reasons, the black holes encountered in Star Trek tend to produce impossible results. The fact that the event horizon is not a tangible object, but rather a mathematical marker that we impose on our description of a black hole to delineate the region inside from that outside, means that the horizon cannot have a "crack," as required by the crew of the *Voyager* when they miraculously escape from a black hole's interior. (Indeed, this notion is so absurd that it makes it onto my ten-best list of Star Trek mistakes described in the last chapter.) And the "quantum singularity life-forms" encountered by the crew of the *Enterprise* as they, and a nearby Romulan Warbird, travel backward and forward in time have a rather unfortunate nesting place for their young: apparently they place them inside natural black holes (which they incorrectly mistake the "artificial" quantum singularity inside the Romulan engine core for). This may be a safe nursery, but it must be difficult to retrieve your children afterward. I remind you that nothing inside a black hole can ever communicate with anything outside one.

Nevertheless, black holes, for all their interesting properties, need not be that exotic. Indeed, the "micro-singularities," presumably small black holes, encountered by shuttlepod one in *Enterprise,* while unexplained, behave more or less like little penetrating meteorites, which would probably not be that far off if such tiny black holes actually existed. In any case, the only black holes we have any evidence for in the universe today result from the collapse of stars much more massive than the Sun. These collapsed objects are so dense that a teaspoon of material inside would weigh many tons. However, it is another remarkable property of black holes that the more massive they are, the less dense they need be when they form. For example, the density of the black hole formed by the collapse of an object 100 million times as massive as our Sun need only be equal to the density of water. An object of larger mass will collapse to form a black hole at a point when it is even less dense. If you keep on extrapolating, you will find that the density required to form a black hole with a mass equal to the mass of the observable universe would be roughly the same as the average density of matter in the universe! We may be living inside a black hole.

In 1974, Stephen Hawking made a remarkable discovery about the nature of black holes. They aren't completely black! Instead, they will emit radiation at a characteristic temperature, which depends on their mass. While the nature of this radiation will give no information whatsoever on what fell into the black hole, the idea that radiation could be emitted from a black hole was nevertheless astounding, and appeared to violate a number of theorems—some of which Hawking had earlier proved—holding that matter could only fall into black holes, not out of them. This remains true, except for the source of the black-hole radiation, which is not normal matter. Instead, it is empty space, which can behave quite exotically—especially in the vicinity of a black hole.

Ever since the laws of quantum mechanics were made consistent with the special theory of relativity, shortly after the Second World War, we have known that empty space is not so empty. It is a boiling, bubbling sea of quantum fluctuations. These fluctuations periodically spit out elementary particle pairs, which exist for time intervals so short that we cannot measure them directly, and then disappear back into the vacuum from which they came. The uncertainty principle of quantum mechanics tells us that there is no way to directly probe empty space over such short time intervals and thus no way to preclude the brief existence of these so-called virtual particles. But although they cannot be measured directly, their presence does affect certain physical processes that we *can* measure, such as the rate and energy of transitions between certain energy levels in atoms. The predicted effect of virtual particles agrees with observations as well as any prediction known in physics.

This brings us back to Hawking's remarkable result about black holes. Under normal circumstances, when a quantum fluctuation creates a virtual particle pair, the pair will annihilate and disappear back into the vacuum in a time short enough so that the violation of conservation of energy (incurred by the pair's creation from nothing) is not observable. However, when a virtual particle pair pops out in the curved space near a black hole, one of the particles may fall into the hole, and then the other can escape and be observed. This is because the particle that falls into the black hole can in principle lose more energy in the process than the amount required to create it from nothing. It thus contributes "negative energy" to the black hole, and the black hole's own energy is therefore decreased. This satisfies the energy-conservation law's balance-sheet, making up for the energy that the escaping particle is observed to have. This is one way of picturing how the black hole emits radiation. Moreover, as the black hole's own energy decreases bit by bit in this process, there is a concomitant decrease in its mass. Eventually, it may completely evaporate, leaving behind only the radiation it produced in its lifetime.

Hawking and many others have gone beyond a consideration of quantum fluctuations of matter in a background curved space to something even more exotic and less well defined. If quantum mechanics applies not merely to matter and radiation but to gravity as well, then on sufficiently small scales quantum fluctuations in spacetime itself must occur. Unfortunately, we have no workable theory for dealing with such processes, but this has not stopped a host of tentative theoretical investigations of phenomena that might result. One of the most interesting speculations is that quantum mechanical processes might allow the spontaneous creation not just of particles but of whole new baby universes. The quantum mechanical formalism describing how this might occur is, at least mathematically, very similar to the wormhole solutions discovered in ordinary general relativity. Via such "Euclidean" wormholes, a temporary "bridge" is created, from which a new universe springs. The possibilities of Euclidean wormhole processes and baby universes are sufficiently exciting that quantum fluctuations were mentioned during Hawking's poker game with Einstein and Newton in the *Next Generation* episode "Descent."[1] If the Star Trek writers were confused, they had a right to be. These issues are unfortunately currently very murky. Until we discover the proper mathematical framework to treat such quantum gravitational processes, all such discussions are shots in the dark.

What is most relevant to us here is not the phenomenon of black-hole evaporation, or even baby universes, as interesting as they may be, but rather the discovery that quantum fluctuations of empty space can, at least in the presence of strong gravitational fields, become endowed with properties reminiscent of those required to hold open a wormhole. The central question, which also has no definitive answer yet, is whether quantum fluctuations near a wormhole can behave sufficiently exotically to allow one to keep a wormhole open.

(By the way, once again, I find the Star Trek writers remarkably prescient in their choice of nomenclature. The Bajoran and Barzan wormholes are said to involve "verteron" fields. I have no idea whether this name was plucked out of a hat or not. However, since virtual particles—the quantum fluctuations in otherwise empty space—are currently the best candidate for Kip Thorne's "exotic matter," I think the Star Trek writers deserve credit for their intuition, if that's what it was.)

More generally, if quantum fluctuations in the vacuum can be exotic, is it possible that some other nonclassical configuration of matter and radiation—like, say, a warp core breach, or perhaps Scotty's "intermix" imbalance in the warp drive—might also fill the bill? Questions such as this remain unanswered. While by no means circumventing the incredible implausibility of stable wormholes in the real universe, they do leave open the larger question of whether wormhole travel is impossible or merely almost impossible. The wormhole issue is not just one of science fact versus science fiction: it is a key that can open doors which many would prefer to leave closed.

Time Machines Revisited. Wormholes, as glorious as they would be for tunneling through vast distances in space, have an even more remarkable potential, glimpsed in the *Voyager* episode "Eye of the Needle." In this episode, the *Voyager* crew discovered a small wormhole leading back to their own "alpha quadrant" of the galaxy. After communicating through it, they found to their horror that it led not to the alpha quadrant they knew and loved but to the alpha quadrant of a generation earlier. The two ends of the wormhole connected space at two different times!

Well, this is another one of those instances in which the *Voyager* writers got it right. If wormholes exist, they can and will be time machines! This startling realization has grown over the last decade, as various theorists, for lack of anything more interesting to do, began to investigate the physics of wormholes a little more seriously. Wormhole time machines are easy to design: perhaps the simplest example (due again to Kip Thorne) is to imagine a wormhole with one end fixed and the other end moving at a fast but sublight speed through a remote region of the galaxy. In principle, this is possible *even if* the length of the wormhole remains unchanged. In my earlier two-dimensional wormhole drawing, just drag the bottom half of the sheet to the left, letting space "slide" past the bottom mouth of the wormhole while this mouth stays fixed relative to the wormhole's other mouth:

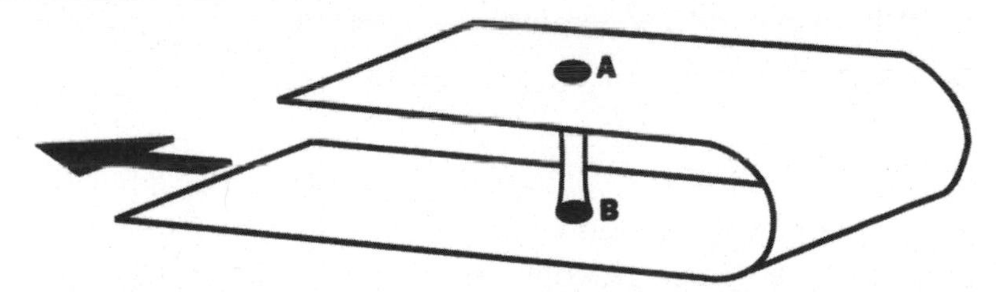

Because the bottom mouth of the wormhole will be moving with respect to the space in which it is situated, while the top mouth will not, special relativity tells us that clocks will tick at different rates at each mouth. On the other hand, if the length of the wormhole remains fixed, then as long as one is inside the wormhole the two ends appear to be at rest relative to each other. In this frame, clocks at either end should be ticking at the same rate. Now slide the bottom sheet back to where it used to be, so that the bottom mouth of the wormhole ends up back where it started relative to the background space. Let's say that this process takes a day, as observed by someone near the bottom mouth. But for an observer near the top mouth, this same process could appear to have taken ten days. If this second observer were to peer through the top mouth to look at the observer located near the bottom mouth, he would see on the wall calendar next to the observer a date nine days earlier! If he now decides to go through the wormhole for a visit, he will travel backward in time. If that doesn't bother you, then you can add the following wrinkle. Say it takes two days to fly through the background space from B to A. Then, if you go through the wormhole, coming out nine days earlier, and hop in a spacecraft, you can arrive back where you started, except you will arrive seven days before you left!

If stable wormholes exist, we must therefore concede that time machines are possible. We now return finally to Einstein's remarks early in the last chapter. Can time travel, and thus stable wormholes, and thus exotic matter with negative energy, be "excluded on physical grounds"?

Wormholes are after all merely one example of time machines that have been proposed in the context of general relativity. Given our previous discussion about the nature of the theory, it is perhaps not so surprising that time travel becomes a possibility. Let's recall the heuristic description of Einstein's equations which I gave earlier:

$$\underset{\{\text{CURVATURE}\}}{\text{Left-hand side}} = \underset{\{\text{MATTER AND ENERGY}\}}{\text{Right-hand side}}$$

The left-hand side of this equation fixes the geometry of spacetime. The right-hand side fixes the matter and energy distribution. Generally we would ask: For a given distribution of matter and energy, what will be the resulting curvature of space? But we can also work backward: For any given geometry of space, including one with "closed timelike curves"—that is, the "causality loops," which allow you to return to where you began in space and time, like the loop the *Enterprise* was caught in before, during, and after crashing into the *Bozeman*—Einstein's equations tell you exactly what distribution of matter and energy must be present. So in principle you can design any kind of time-travel universe you want; Einstein's equations will tell you what matter and energy distribution is necessary. The key question then simply becomes: Is such a matter and energy distribution physically possible?

We have already seen how this question arises in the context of wormholes. Stable wormholes require exotic matter with negative energy. Kurt Gödel's time-machine solution in general relativity involves a universe with constant uniform energy density and zero pressure which spins but does not expand. More recently, a proposed time machine involving "cosmic strings" was shown to require a negative-energy configuration. In fact, it was recently proved that any configuration of matter in general relativity which might allow time travel must involve exotic types of matter with negative energy as viewed by at least one observer.

It is interesting that almost all the episodes in Star Trek involving time travel or temporal distortions also involve some catastrophic form of energy release, usually associated with a warp core breach. For example, the temporal causality loop in which the *Enterprise* was trapped resulted only after (although the concepts of "before" and "after" lose their meaning in a causality loop) a collision with the *Bozeman,* which caused the warp core to breach and thereby caused the destruction of the *Enterprise,* a series of events that kept repeating over and over, until finally in one cycle the crew managed to avoid the collision. The momentary freezing of time aboard the *Enterprise,* discovered by Picard, Data, Troi, and LaForge in the episode "Timescape," also appears to have been produced by a nascent warp core breach combined with a failure of the engine core aboard a nearby Romulan vessel. In "Time Squared," a vast "energy vortex" propelled Picard back in time. In the original example of Star Trek time travel, "The Naked Time," the *Enterprise* was thrown back three days following a warp core implosion. And the mammoth space-time distortion in the final episode of *The Next Generation,* which travels backward in time and threatens to engulf the entire universe, was caused by the simultaneous explosion of three different temporal versions of the *Enterprise,* which converged at the same point in space. And in the series *Enterprise,* during the Temporal Cold War, the spacetime distortion caused by the 31st century time-traveler Crewman Daniels as he transports Captain Archer and himself back to 1944 is sufficiently violent to lead to Daniels's death—that is until a new timeline is restored and Daniels heads back to the future.

So, time travel in the real universe, as in the Star Trek universe, seems to hinge on the possibility of exotic configurations of matter. Could some sufficiently advanced alien civilization construct a stable wormhole? Or can we characterize *all* mass distributions that might lead to time travel and then exclude them, as a set, "on physical grounds," as Einstein might have wished? To date, we do not know the answer. Some specific time machines—such as Gödel's, and the cosmic-string-based system—have been shown to be unphysical. While wormhole time travel has yet to be definitively ruled out, preliminary investigations suggest that the quantum gravitational fluctuations themselves may cause wormholes to self-destruct before they could lead to time travel.

Until we have a theory of quantum gravity, the final resolution of the issue of time travel is likely to remain unresolved. Nevertheless, several brave individuals, including Stephen Hawking, have already tipped their hand. Hawking is convinced that time machines are impossible, because of the obvious paradoxes that might result, and he has proposed a "chronology-protection conjecture," to wit: "The laws of physics do not allow the appearance of closed time-like curves." Or perhaps, if Hawking is wrong, and the laws of physics do not rule out changing chronologies, maybe we can count on laws, like the future Temporal Accord revealed in the series *Enterprise*, made by intelligent species so as not to interfere in the past.

I am personally inclined to agree with Hawking in this case. Nevertheless, physics is not done by fiat. As I have stated earlier, general relativity often outwits our naive expectations. As a warning, I provide two historical precedents. Twice before (that I know of), eminent theorists have argued that a proposed phenomenon in general relativity should be dismissed because the laws of physics must forbid it:

1. When the young astrophysicist Subrahmanyan Chandrasekhar proposed that stellar cores more massive than 1.4 times the mass of the Sun cannot, after burning all their nuclear fuel, settle down as white dwarfs but must continue to collapse due to gravity, the eminent physicist Sir Arthur Eddington dismissed the result in public, stating, "Various accidents may intervene to save the star, but I want more protection than that. I think there should be a law of nature to prevent a star from behaving in this absurd way!" At the time, much of the astrophysics community sided with Eddington. A half century later, Chandrasekhar shared the Nobel Prize for his insights, which have long since been verified.

2. Slightly over 20 years after Eddington dismissed Chandrasekhar's claim, a remarkably similar event occurred at a

conference in Brussels. J. Robert Oppenheimer, the distinguished American theoretical physicist and father of the atomic bomb, had calculated that objects called neutron stars—left over after supernovae and even more dense than white dwarfs—could not be larger than about twice the mass of the Sun without collapsing further to form what we would now call a black hole. The equally distinguished John Archibald Wheeler argued that this result was impossible, for precisely the reason Eddington had given for his earlier rejection of Chandrasekhar's claim: somehow the laws of physics must protect objects from such an absurd fate. Within a decade, Wheeler would completely capitulate and, ironically, would become known as the man who gave black holes their name.

4

Data Ends the Game

For I dipt into the future, far as human eye
could see,
Saw the Vision of the world, and all the
wonder that would be.

—From "Locksley Hall," by Alfred Lord Tennyson
(posted aboard the starship *Voyager*)

Whether or not the Star Trek future can include a stable wormhole, and whether or not the *Enterprise* crew could travel back in time to nineteenth-century San Francisco, the real stakes in this cosmic poker game derive from one of the questions that led us to discuss curved spacetime in the first place: Is warp drive possible? For, barring the unlikely possibility that our galaxy is riddled with stable wormholes, it is abundantly clear from our earlier discussions that without something like it, most of the galaxy will always remain beyond our reach. It is finally time to address this vexing question. The answer is a resounding "Maybe!"

Once again we are guided by the linguistic perspicacity of the Star Trek writers. I have described how no rocket propulsion mechanism can ever get around the three roadblocks to interstellar travel set up by special relativity: First, nothing can travel faster than the speed of light in empty space. Second, objects that travel near the speed of light will have clocks that are slowed down. Third, even if a rocket could accelerate a spacecraft to near the speed of light, the fuel requirements would be prohibitive.

The idea is not to use any sort of rocket at all for propulsion, but instead to use spacetime itself—by warping it. General relativity requires us to be a little more precise in our statements about motion. Instead of saying that nothing can travel faster than the speed of light, we must state that nothing can travel *locally* any faster than the speed of light. This means that nothing can travel faster than the speed of light *with respect to local distance markers.* However, if spacetime is curved, local distance markers need not be global ones.

Let me use the universe itself as an example. Special relativity tells me that all observers who are at rest with respect to their local surroundings will have clocks that tick at the same rate. Thus, as I move throughout the universe, I can periodically stop and place clocks at regular intervals in space and expect that they will all keep the same time. General relativity does not change this result. Clocks that are locally at rest will all keep the same time. However, general relativity allows spacetime itself to expand. Objects on opposite sides of the observable universe are flying apart at almost the speed of light, yet they remain at rest relative to their local surroundings. In fact, if the universe is expanding uniformly and if it is large enough—both of which appear to be the case—there exist objects we cannot yet see which are at this very moment moving away from us far faster than the speed of light, even though any civilizations in these far reaches of the universe can be locally at rest with respect to their surroundings.

The curvature of space therefore produces a loophole in special relativistic arguments—a loophole large enough to drive a Federation starship through. If spacetime itself can be manipulated, objects can travel locally at very slow velocities, yet an accompanying expansion or contraction of space could allow huge distances to be traversed in short time intervals. We have already seen how an extreme manipulation—namely, cutting and pasting distant parts of the universe together with a wormhole—might create shortcuts through spacetime. What is argued here is that even if we do not resort to this surgery, faster-than-light travel might globally be possible, even if it is not locally possible.

A proof in principle of this idea was developed by a physicist then in Wales, Miguel Alcubierre, who for fun decided to explore whether a consistent solution in general relativity could be derived which would correspond to "warp travel." He was able to demonstrate that it was possible to tailor a spacetime configuration wherein a spacecraft could travel between two points in an arbitrarily short time. Moreover, throughout the journey the spacecraft could be moving with respect to its local surroundings at speeds much less than the speed of light, so that clocks aboard the spacecraft would remain synchronized with those at its place of origin and at its destination. General relativity appears to allow us to have our cake and eat it too.

The idea is straightforward. If spacetime can locally be warped so that it expands behind a starship and contracts in front of it, then the craft will be propelled along with the space it is in, like a surfboard on a wave. The craft will never travel locally faster than the speed of light, because the light, too, will be carried along with the expanding wave of space.

One way to picture what is happening is to imagine yourself on the starship. If space suddenly expands behind you by a huge amount, you will find that the starbase you just left a few minutes ago is now many light-years away. Similarly, if space contracts in front of you, you will find that the starbase you are heading for, which formerly was a few light-years away, is now close to you, within reach by normal rocket propulsion in a matter of minutes.

It is also possible to arrange the geometry of spacetime in this solution so that the huge gravitational fields necessary to expand and contract space in this way are never large near the ship or any of the starbases. In the vicinity of the ship and the bases, space can be almost flat, and therefore clocks on the ship and the starbases remain synchronized. Somewhere in between the ship and the bases, the tidal forces due to gravity will be immense, but that's OK as long as we aren't located there.

This scenario must be what the Star Trek writers intended when they invented warp drive, even if it bears little resemblance to the technical descriptions they have provided. It fulfills all the requirements we listed earlier for successful controlled intergalactic space travel: (1) faster-than-light travel, (2) no time dilation, and (3) no resort to rocket propulsion. Of course, we have begged a pretty big question thus far. By making spacetime itself dynamical, general relativity allows the creation of "designer spacetimes," in which almost any type of motion in space and time is possible. However, the cost is that the theory relates these spacetimes to some underlying distribution of matter and energy. Thus, for the desired spacetime to be "physical," the underlying distribution of matter and energy must be attainable. I will return to this question shortly.

First, however, the wonder of such "designer spacetimes" is that they allow us to return to Newton's original challenge and to create inertial dampers and tractor beams. The idea is identical to warp drive. If spacetime around the ship can be warped, then objects can move apart or together without experiencing any sense of local acceleration, which you will recall was Newton's bane. To avoid the incredible accelerations required to get to impulse sublight speeds, one must resort to the same spacetime shenanigans as one does to travel at warp speeds. The distinction between impulse drive and warp drive is thus diminished. Similarly, to use a tractor beam to pull a heavy object like a planet, one merely has to expand space on the other side of the planet and contract it on the near side. Simple!

Warping space has other advantages as well. Clearly, if spacetime becomes strongly curved in front of the *Enterprise*, then any light ray—or phaser beam, for that matter—will be deflected away from the ship. This is doubtless the principle behind deflector shields. Indeed, we are told that the deflector shields operate by "coherent graviton emission." Since gravitons are by definition particles that transmit the force of gravity, then "coherent graviton emission" is nothing other than the creation of a coherent gravitational field. A coherent gravitational field is, in modern parlance, precisely what curves space! So once again the Star Trek writers have at least settled upon the right language.

I would imagine that the Romulans' cloaking device might operate in a similar manner. In fact, an *Enterprise* that has its deflector shield deployed should be very close to a cloaked *Enterprise*. After all, the reason we see something that doesn't shine of its own accord is that it reflects light, which travels back to us. Cloaking must somehow warp space so that incident light rays bend around a Warbird instead of being reflected from it. (Interestingly, several U.S. investigators have been investigating a kind of cloaking device for aircraft, at least cloaking from radar. The device reflects and bends incoming radio waves around the craft, not by using gravity, but by using electromagnetic interactions. Such a thing is possible if one is aiming at manipulating one specific kind of radiation, like radio waves. Alas, to bend all kinds of radiation equally, one has to resort to gravity.) The distinction between this and deflecting light rays away from the *Enterprise* is thus pretty subtle. In this connection, a question that puzzled many trekkers until the *Next Generation* episode "The Pegasus" aired was, Why didn't the Federation employ cloaking technology? It would certainly seem, in light of the above, that any civilization that could develop deflector shields could develop cloaking devices. And as we learned in "The Pegasus," the Federation was limited in its development of cloaking devices by treaty rather than by technology. (Indeed, as became evident in "All Good Things . . ," the last episode of *Next Generation,* the Federation eventually seems to have allowed cloaking on starships.) And clearly the Coalition of Planets was able to manipulate gravity even as early as the first *Enterprise*, which not only was warp capable, but also had artificial gravity generators producing a "sweet spot" on each ship where gravity allows you to sit on the ceiling, at least according to Ensign Mayweather.

Finally, given this general-relativistic picture of warp drive, warp speeds take on a somewhat more concrete meaning. The warp speed would be correlated to the contraction and expansion factor of the spatial volume in front of and behind the ship. Warp-speed conventions have never been particularly stable: between the first and second series, Gene Roddenberry apparently decided that warp speeds should be recalibrated so that nothing could exceed warp 10. This meant that warp speed could not be a simple logarithmic scale, with, say, warp 10 being 2^{10} = 1024 x light speed. According to the *Next Generation Technical Manual,* warp 9.6, which is the highest normal rated speed for the *Enterprise-D,* is 1909 x the speed of light, and warp 10 is infinite. It is interesting to note that in spite of this recalibration, objects (such as the Borg cube) are periodically sighted that go faster than warp 10, so I suppose one shouldn't concern oneself unduly about understanding the details.

Well, so much for the good news . . .

Having bought into warp drive as a nonimpossibility (at least in principle), we finally have to face up to the consequences for the right-hand side of Einstein's equations—namely, for the distribution of matter and energy required to produce the requisite curvature of spacetime. And guess what? The situation is almost *worse* than it was for wormholes. Observers traveling at high speed through a wormhole can measure a negative energy. For the kind of matter needed to produce a warp drive, even an observer at rest with respect to the starship—that is, someone on board—can measure a negative energy.

This result is not too surprising. At some level, the exotic solutions of general relativity required to keep wormholes open, allow time travel, and make warp drive possible all imply that on some scales matter must gravitationally repel other matter. There is a theorem in general relativity that this condition is generally equivalent to requiring the energy of matter to be negative for some observers.

What *is* surprising, perhaps, is the fact, mentioned earlier, that quantum mechanics, when combined with special relativity, implies that at least on microscopic scales the local distribution of energy can be negative. Indeed, as I noted in chapter 3, quantum fluctuations often have this property. The key question, which remains unanswered to date, is whether the laws of physics as we know them will allow matter to have this property on a macroscopic scale. It is certainly true that currently we haven't the slightest idea of how one could create such matter in any physically realistic way.

Interestingly it has just been realized, in fact since the first edition of this book appeared, that nature itself has endowed empty space with a kind of energy that is gravitationally repulsive. It turns out that the dominant energy in the universe appears to be the energy of empty space, about three times more significant than all the energy contained in all the galaxies and clusters in the universe. This "dark energy," as it has become known, is causing the expansion of the universe to speed up, just as the famous "cosmological constant" that Einstein invented and then discarded a few years later would do. We currently have no idea whatsoever why this dark energy is out there, or what its origin is. However, we do know that if it is really the energy of empty space, which quantum mechanics allows to be non-zero, then there is no way it could be harnessed to do any useful work like transporting spacecraft around.

However, ignore for the moment the potential obstacles to creating such material, and suppose that it will someday be possible to create exotic matter by using some sophisticated quantum mechanical engineering of matter or of empty space. Even so, the energy requirements to do any of the remarkable playing around with spacetime described here would likely make the power requirement for accelerating to impulse speed seem puny. Consider the mass of the Sun, which is about a million times the mass of the Earth. The gravitational field at the surface of the Sun is sufficient to bend light by less than 1/1000 of a degree. Imagine the extreme gravitational fields that would have to be generated near a starship to deflect an oncoming phaser beam by 90°! (This is one of the many reasons why the famous "slingshot effect"—first used in the classic episode "Tomorrow Is Yesterday" to propel the *Enterprise* backward in time, again in *Star Trek IV: The Voyage Home,* and also mentioned in the *Next Generation* episode "Time Squared"—is completely impossible. The gravitational field near the surface of the Sun is minuscule in terms of the kind of gravitational effects required to perturb spacetime in the ways we have discussed here.) One way to estimate how much energy would have to be generated is to imagine producing a black hole the size of the *Enterprise*—since certainly a black hole of this size would produce a gravitational field that could significantly bend any light beam that traveled near it. The mass of such a black hole would be about 10 percent of the mass of the Sun. Expressed in energy units, it would take more than the total energy produced by the Sun during its entire lifetime to generate such a black hole.

So where do we stand at the end of this game? We know enough about the nature of spacetime to describe explicitly how one might, at least in principle, utilize curved space to achieve many of the essentials of interstellar space travel à la Star Trek. We know that without such exotic possibilities we will probably never voyage throughout the galaxy on round trips. On the other hand, we have no idea whether the *physical conditions* needed to achieve any of these things are realizable in practice or even allowed in principle. Finally, even if they were, it is clear that any civilization putting these principles into practice would have to harness energies vastly in excess of anything imaginable today.

I suppose one might take the optimistic view that these truly remarkable wonders are at least not *a priori* impossible. They merely hinge on one remote possibility: the ability to create and sustain exotic matter and energy. There is reason for hope, but I must admit that I remain skeptical. Like my colleague Stephen Hawking, I believe that the paradoxes involved in round-trip time travel are likely to rule it out for any sensible physical theory. Since virtually the same conditions of energy and matter are required for warp travel and deflector shields, I'm not anticipating them either—though I have been wrong before.

Nevertheless, I am still optimistic. What to me is really worth celebrating is the remarkable body of knowledge that has brought us to this fascinating threshold. We live in a remote corner of one of 100 billion galaxies in the observable universe. And like insects on a rubber sheet, we live in a universe whose true form is hidden from direct view. Yet in the course of less than twenty generations—from Newton to today—we have utilized the simple laws of physics to illuminate the depths of space and time. It is likely that we may never be able to board ships headed for the stars, but even imprisoned on this tiny blue planet we have been able to penetrate the night sky to reveal remarkable wonders, and there is no doubt more to come. If physics cannot give us what we need to roam the galaxy, it is giving us what we need to bring the galaxy to us.

SECTION TWO

MATTER MATTER EVERYWHERE

In which the reader explores transporter beams, warp drives, dilithium crystals, matter-antimatter engines, and the holodeck

5

Atoms or Bits

"Reg, transporting *really is* the safest way to travel."

—Geordi LaForge to Lieutenant Reginald Barclay, in "Realm of Fear"

Life imitates art. Lately, I keep hearing the same question: "Atoms or bits—where does the future lie?" Thirty years ago, Gene Roddenberry dealt with this same speculation, driven by another imperative. He had a beautiful design for a starship, with one small problem: like a penguin in the water, the *Enterprise* could glide smoothly through the depths of space, but like a penguin on the ground it clearly would have trouble with its footing if it ever tried to land. More important perhaps, the meager budget for a weekly television show precluded landing a huge starship every week.

How then to solve this problem? Simple: make sure the ship would never need to land. Find some other way to get the crew members from the ship to a planet's surface. No sooner could you say, "Beam me up" than the transporter was born.

Perhaps no other piece of technology, save for the warp drive, so colors every mission of every starship of the Federation. And even those who have never watched a Star Trek episode recognize the magic phrase on the preceding page. It has permeated our popular culture. I recently heard about a young man who, while inebriated, drove through a red light and ran into a police cruiser that happened to be lawfully proceeding through the intersection. At his hearing, he was asked if he had anything to say. In well-founded desperation, he replied, "Yes, your honor," stood up, took out his wallet, flipped it open, and muttered into it, "Beam me up, Scotty!"

The story is probably apochryphal, but it is testimony to the impact that this hypothetical technology has had on our culture—an impact all the more remarkable given that probably no single piece of science fiction technology aboard the *Enterprise* is so utterly implausible. More problems of practicality and principle would have to be overcome to create such a device than you might imagine. The challenges involve the whole spectrum of physics and mathematics, including information theory, quantum mechanics, Einstein's relation between mass and energy, elementary particle physics, and more.

Which brings me to the atoms versus bits debate. The key question the transporter forces us to address is the following: Faced with the task of moving, from the ship to a planet's surface, roughly 10^{28} (1 followed by 28 zeroes) atoms of matter combined in a complex pattern to make up an individual human being, what is the fastest and most efficient way to do it? This is a very timely question, because we are facing exactly the same quandary as we consider how best to disseminate the complex pattern of roughly 10^{26} atoms in an average paperback book. When the first edition of this book was written, it was just a vague idea, but now it is commonplace to find books in which the atoms themselves are secondary. What matters more are the bits.

Consider, for example, a library book. A library buys one copy—or, for some lucky authors, several copies—of a book, which it stores and lends out for use by one individual at a time. However, in a digital library the same information can be stored as bits. A bit is a 1 or a 0, which is combined in groups of eight, called bytes, to represent words or numbers. This information is stored in the memory cores of computers, in which each bit is represented as either a magnetized or charged (1) or unmagnetized or uncharged (0) region. Now an arbitrarily large number of users can access the same memory location on a computer at essentially the same time, so in a digital library every single person on Earth who might otherwise have to buy a book can read it from a single source. Clearly, in this case, having on hand the actual atoms that make up the book is less significant, and certainly less efficient, than storing the bits (although it will play havoc with authors' royalties).

So, what about people? If you are going to move people around, do you have to move their atoms or just their information? At first you might think that moving the information is a lot easier; for one thing, information can travel at the speed of light. However, in the case of people, you have two problems you don't have with books: first, you have to extract the information, which is not so easy, and then you have to recombine it with matter. After all, people, unlike books, require the atoms.

The Star Trek writers seem never to have got it exactly clear what they want the transporter to do. Does the transporter send the atoms *and* the bits, or just the bits? You might wonder why I make this point, since the *Next Generation Technical Manual* describes the process in detail: First the transporter locks on target. Then it scans the image to be transported, "dematerializes" it, holds it in a "pattern buffer" for a while, and then transmits the "matter stream," in an "annular confinement beam," to its destination. The transporter thus apparently sends out the matter along with the information.

The only problem with this picture is that it is inconsistent with what the transporter sometimes does. On at least two well-known occasions, the transporter has started with one person and beamed up two. In the famous classic episode "The Enemy Within," a transporter malfunction splits Kirk into two different versions of himself, one good and one evil. In a more interesting, and permanent, twist, in the *Next Generation* episode "Second Chances," we find out that Lieutenant Riker was earlier split into two copies during transportation from the planet Nervala IV to the *Potemkin*. One version returned safely to the *Potemkin* and one was reflected back to the planet, where he lived alone for eight years.

If the transporter carries both the matter stream and the information signal, this splitting phenomenon is impossible. The number of atoms you end up with has to be the same as the number you began with. There is no possible way to replicate people in this manner. On the other hand, if only the information were beamed up, one could imagine combining it with atoms that might be stored aboard a starship and making as many copies as you wanted of an individual.

A similar problem concerning the matter stream faces us when we consider the fate of objects beamed out into space as "pure energy." For example, in the *Next Generation* episode "Lonely among Us," Picard chooses at one point to beam out as pure energy, free from the constraints of matter. After this proves a dismal and dangerous experience, he manages to be retrieved, and his corporeal form is restored from the pattern buffer. But if the matter stream had been sent out into space, there would have been nothing to restore at the end.

So, the Star Trek manual notwithstanding, I want to take an agnostic viewpoint here and instead explore the myriad problems and challenges associated with each possibility: transporting the atoms or the bits.

⋐⋑

When a Body Has No Body. Perhaps the most fascinating question about beaming—one that is usually not even addressed—

is, What comprises a human being? Are we merely the sum of all our atoms? More precisely, if I were to re-create each atom in your body, in precisely the same chemical state of excitation as your atoms are in at this moment, would I produce a functionally identical person who has exactly all your memories, hopes, dreams, spirit? There is every reason to expect that this would be the case, but it is worth noting that it flies in the face of a great deal of spiritual belief about the existence of a "soul" that is somehow distinct from one's body. What happens when you die, after all? Don't many religions hold that the "soul" can exist after death? What then happens to the soul during the transport process? In this sense, the transporter would be a wonderful experiment in spirituality. If a person were beamed aboard the *Enterprise* and remained intact and observably unchanged, it would provide dramatic evidence that a human being is no more than the sum of his or her parts, and the demonstration would directly confront a wealth of spiritual beliefs.

For obvious reasons, this issue is studiously avoided in Star Trek. However, in spite of the purely physical nature of the dematerialization and transport process, the notion that some nebulous "life force" exists beyond the confines of the body is a constant theme in the series. The entire premise of the second and third Star Trek movies, *The Wrath of Khan* and *The Search for Spock,* is that Spock, at least, has a "katra"—a living spirit—which can exist apart from the body. More recently, in the *Voyager* series episode "Cathexis," the "neural energy"—akin to a life force—of Chakotay is removed and wanders around the ship from person to person in an effort to get back "home."

I don't think you can have it both ways. Either the "soul," the "katra," the "life force," or whatever you want to call it is part of the body, and we are no more than our material being, or it isn't. In an effort not to offend religious sensibilities, even a Vulcan's, I will remain neutral in this debate. Nevertheless, I thought it worth pointing out before we forge ahead that even the basic premise of the transporter—that the atoms *and* the bits are all there is—should not be taken lightly.

The Problem with Bits. Many of the problems I will soon discuss could be avoided if one were to give up the requirement of transporting the atoms along with the information. After all, anyone with access to the Internet knows how easy it is to transport a data stream containing, say, the detailed plans for a new car, along with photographs. Moving the actual car around, however, is nowhere near as easy. Nevertheless, two rather formidable problems arise even in transporting the bits. The first is a familiar quandary, faced, for example, by the last people to see Jimmy Hoffa alive: How are we to dispose of the body? If just the information is to be transported, then the atoms at the point of origin must be dispensed with and a new set collected at the reception point. This problem is quite severe. If you want to zap 10^{28} atoms, you have quite a challenge on your hands. Say, for example, that you simply want to turn all this material into pure energy. How much energy would result? Well, Einstein's formula $E = mc^2$ tells us. If one suddenly transformed 50 kilograms (a light adult) of material into energy, one would release the energy equivalent of somewhere in excess of a thousand 1-megaton hydrogen bombs. It is hard to imagine how to do this in an environmentally friendly fashion.

There is, of course, another problem with this procedure. If it is possible, then replicating people would be trivial. Indeed, it would be much easier than transporting them, since the destruction of the original subject would then not be necessary. Replication of inanimate objects in this manner is something one can live with, and indeed the crew members aboard starships do seem to live with this. However, replicating living human beings would certainly be cause for trouble (à la Riker in "Second Chances"). Indeed, if recombinant DNA research today has raised a host of ethical issues, the mind boggles at those that would be raised if complete individuals, including memory and personality, could be replicated at will. People would be like computer programs, or drafts of a book kept on disk. If one of them gets damaged or has a bug, you could simply call up a backup version.

Ok, Keep the Atoms. The preceding arguments suggest that on both practical and ethical grounds it might be better to imagine a transporter that carries a matter stream along with the signal, just as we are told the Star Trek transporters do. The problem then becomes, How do you move the atoms? Again, the challenge turns out to be energetics, although in a somewhat more subtle way.

What would be required to "dematerialize" something in the transporter? To answer this, we have to consider a little more carefully a simpler question: What is matter? All normal matter is made up of atoms, which are in turn made up of very dense central nuclei surrounded by a cloud of electrons. As you may recall from high school chemistry or physics, most of the volume of an atom is empty space. The region occupied by the outer electrons is about ten thousand times larger than the region occupied by the nucleus.

Why, if atoms are mostly empty space, doesn't matter pass through other matter? The answer to this is that what makes a wall solid is not the existence of the particles but of the electric fields between the particles. My hand is stopped from going through my desk when I slam it down primarily because of the electric repulsion felt by the electrons in the atoms in my hand due to the presence of the electrons in the atoms of the desk and *not* because of the lack of available space for the electrons to move through.

These electric fields not only make matter corporeal, in the sense of stopping objects from passing through one another, but they also hold the matter together. To alter this normal situation, one must therefore overcome the electric forces between atoms. Overcoming these forces will require work, which takes energy. Indeed, this is how all chemical reactions work. The configuration of individual sets of atoms and their binding to one another are altered through the exchange of energy. For example, if one injects some energy into a mixture of ammonium nitrate and fuel oil, the molecules of the two materials can rearrange, and in the process the "binding energy" holding the original materials can be released. This release, if fast enough, will cause a large explosion.

The binding energy between atoms is, however, minuscule compared to the binding energy of the particles—protons and neutrons—that make up the incredibly dense nuclei of atoms. The forces holding these particles together in a nucleus result in binding energies that are millions of times stronger than the atomic binding energies. Nuclear reactions therefore release significantly more energy than chemical reactions, which is why nuclear weapons are so powerful.

Finally, the binding energy that holds together the elementary particles, called quarks, which make up the protons and neutrons themselves is yet larger than that holding together the protons and neutrons in nuclei. In fact, it is currently believed—based on all calculations we can perform with the theory describing the interactions of quarks—that it would take an infinite amount of energy to completely separate the quarks making up each proton or neutron.

Based on this argument, you might expect that breaking matter completely apart into quarks, its fundamental constituents, would be impossible—and it is, at least at room temperature. However, the same theory that describes the interactions of quarks inside protons and neutrons tells us that if we were to heat up the nuclei to about 1000 billion degrees (about a million times hotter than the temperature at the core of the Sun), then not only would the quarks inside lose their binding energies but at around this temperature matter will suddenly lose almost all of its mass. Matter will turn into radiation—or, in the language of our transporter, matter will dematerialize.

So, all you have to do to overcome the binding energy of matter at its most fundamental level (indeed, at the level referred to in the Star Trek technical manual) is to heat it up to 1000 billion degrees. In energy units, this implies providing about 10 percent of the rest mass of protons and neutrons in the form of heat. To heat up a sample the size of a human being to this level would require, therefore, about 10 percent of the energy needed to annihilate the material—or the energy equivalent of a hundred 1-megaton hydrogen bombs.

One might suggest, given this daunting requirement, that the scenario I have just described is overkill. Perhaps we don't have to break down matter to the quark level. Perhaps a dematerialization at the proton and neutron level, or maybe even the atomic level, is sufficient for the purposes of the transporter. Certainly the energy requirements in this case would be vastly less, even if formidable. Unfortunately, hiding this problem under the rug exposes one that is more severe. For once you have the matter stream, made now of individual protons and neutrons and electrons, or perhaps whole atoms, you have to transport it—presumably at a significant fraction of the speed of light.

Now, in order to get particles like protons and neutrons to move near the speed of light, one must give them an energy comparable to their rest-mass energy. This turns out to be about ten times larger than the amount of energy required to heat up and "dissolve" the protons into quarks. Nevertheless, even though it takes more energy per particle to accelerate the protons to near light speed, this is still easier to accomplish than to deposit and store enough energy inside the protons for long enough to heat them up and dissolve them into quarks. This is why today we can build, albeit at great cost, enormous particle accelerators—like Fermilab's Tevatron, in Batavia, Illinois—that can accelerate individual protons up to more than 99.9 percent of the speed of light, but we have not yet managed to build an accelerator that can bombard protons with enough energy to "melt" them into their constituent quarks. In fact, it is one of the goals of physicists designing the next generation of large accelerators—including one device being built at Brookhaven National Laboratory, on Long Island—to actually achieve this "melting" of matter.

Yet again I am impressed with the apt choice of terminology by the Star Trek writers. The melting of protons into quarks is what we call in physics a phase transition. And lo and behold, if one scours the *Next Generation Technical Manual* for the name of the transporter instruments that dematerialize matter, one finds that they are called "phase transition coils."

So the future designers of transporters will have a choice. Either they must find an energy source that will temporarily produce a power that exceeds the total power consumed on the entire Earth today by a factor of about 10,000, in which case they could make an atomic "matter stream" capable of moving along with the information at near the speed of light, or they could reduce the total energy requirements by a factor of 10 and discover a way to heat up a human being instantaneously to roughly a million times the temperature at the center of the Sun.

If This Is the Information Superhighway, We'd Better Get in the Fast Lane. As I write this on my home computer, I marvel at the speed with which this technology has developed since I bought my first Macintosh a little over two decades ago. I remember that the internal memory in that machine was 128 kilobytes, as opposed to the 512 megabytes in my current machine and the 2048 megabytes in the fast workstation I have in my office in Case Western Reserve's Physics Department. Thus, in two decades my computer internal-memory capabilities have increased by a factor of 10,000! This increase has been matched by an increase in the capacity of my hard-drive memory. My first machine had no hard drive at all and thus had to work from floppy disks, which held 400 kilobytes of information. My present home machine has a 500-gigabyte hard drive—an increase of more than a factor of 1 million in my storage capabilities. The speed of my home system has also greatly increased in the last two decades. For doing actual detailed numerical calculations, I estimate that my present machine is over 1,000 times faster than my first Macintosh. My office workstation is perhaps ten times faster still, performing well over 1 billion instructions per second!

Even at the cutting edge, the improvement has been impressive. The fastest computers used for general-purpose computing have increased in speed and memory capability by a factor of about 100 in the past decade. And I am not including here computers built for special-purpose work: these little marvels can have effective speeds exceeding thousands of billions of instructions per second. In fact, it has been shown that in principle certain special-purpose devices might be built using biological, DNA-based systems, which could be orders of magnitude faster. In fact, research is now ongoing on a type of computer called a quantum computer, about which I will have more to say later. It would utilize the weird properties of quantum mechanics to perform literally gazillions of computations at the same time. It is not at all clear whether such a device will ever be practical, but the fact that it is possible, even in principle, has legions of physicists excited.

One might wonder where all this is heading, and whether we can extrapolate the past rapid growth to the future. Another valid question is whether we need to keep up this pace. I find already that the rate-determining step in the information superhighway is the end user. We can assimilate only so much information. Try surfing the Internet for a few hours, if you want a graphic example of this. I often wonder why, with the incredible power at my disposal, my own productivity has not increased nearly as dramatically as my computer's. Indeed, it has perhaps been reduced! I think the answer is clear. I am not limited by my computer's capabilities but by my own capabilities. It has been argued that for this reason computing machines could be the next phase of human evolution. It is certainly true that Data, even without emotions, is far superior to his human crewmates in most respects. And, as determined in "The Measure of a Man," he is a genuine life-form.

But I digress. The point of noting the growth of computer capability in the last decade is to consider how it compares with what we would need to handle the information storage and retrieval associated with the transporter. And of course, it doesn't come anywhere close.

Let's make a simple estimate of how much information is encoded in a human body. Start with our standard estimate of 10^{28} atoms. For each atom, we first must encode its location, which requires three coordinates (the x, y, and z positions). Next, we would have to record the internal state of each atom, which would include things like which energy levels are occupied by its electrons, whether it is bound to a nearby atom to make up a molecule, whether the molecule is vibrating or rotating, and so forth. Let's be conservative and assume that we can encode all the relevant information in a kilobyte of data. (This is roughly the amount of information on a double-spaced typewritten page.) That means we would need roughly 10^{28} kilobytes to store a human pattern in the pattern buffer. I remind you that this is a 1 followed by 28 zeros.

After making this estimate in the first edition of this book, I received a host of letters from people who proposed ways to reduce this number. For example, they would say, everyone has a heart, two lungs, etc., so we don't need to record all of that data; or they would say, well, each individual is determined by their DNA, so just store the information in the DNA. There are, of course, concerns about each of these proposals. For example, clearly our physical condition is not uniquely determined by our DNA. It also depends upon how much exercise we get, how much we eat, whether we smoke, etc. But even ignoring all of these issues, clearly what really matters is our brains. They make us who we really are, and to reproduce them exactly at the molecular level, where our memories and personality are probably stored, would require us to store, within a few orders of magnitude, the amount of data I estimated above.

Compare this with, say, the total information stored in all the books ever written. The largest libraries contain several million volumes, so let's be very generous and say that there are a billion different books in existence (one written for every five people now alive on the planet). Say each book contains the equivalent of a thousand typewritten pages of information (again on the generous side)—or about a megabyte. Then all the information in all the books ever written would require about 10^{12}, or about a million million, kilobytes of storage. This is about sixteen orders of magnitude—or about one ten-millionth of a billionth—smaller than the storage capacity needed to record a single human pattern! When numbers get this large, it is difficult to comprehend the enormity of the task. Perhaps a comparison is in order. The storage requirements for a human pattern are ten thousand times as large, compared to the information in all the books ever written, as the information in all the books ever written is compared to the information on this page.

Storing this much information is, in an understatement physicists love to use, nontrivial. At present, the largest commercially available single hard disks store about 1,000 gigabytes, or 1 million megabytes, of information. If each disk is about 10 cm thick, then if we stacked all the disks currently needed to store a human pattern on top of one another, they would reach 1/300 of the way to the center of the galaxy—about 100 light-years, or about one weeks' travel in the *Enterprise* at warp 9!

Retrieving this information in real time is no less of a challenge. The fastest digital information transfer mechanisms at present can move somewhat less than about 10 gigabytes per second. At this rate, it would take about 20 times the present age of the universe (assuming an approximate age of 10 billion years) to write the data describing a human pattern to tape! Imagine than the dramatic tension: Kirk and McCoy have escaped to the surface of the penal colony at Rura Penthe. You don't have even the age of the universe to beam them back, but rather just seconds to transfer a million billion billion megabytes of information in the time it takes the jailor to aim his weapon before firing.

I think the point is clear. This task dwarfs the ongoing Human Genome Project, which scanned and essentially recorded the complete human genetic code contained in microscopic strands of human DNA. This was a multibillion-dollar endeavor, carried out over at least a decade and requiring dedicated resources in many laboratories around the world. So you might imagine that I am mentioning it simply to add to the transporter-implausibility checklist. However, while the challenge is daunting, I think this is one area that could possibly be up to snuff in the twenty-third century. My optimism stems merely from extrapolating the present growth rate of computer technology. Using my previous yardstick of improvement in storage and speed by a factor of 100 each decade, and dividing it by 10 to be conservative—and given that we are about 19 powers of 10 short of the mark now—one might expect that 190 years from now, at the dawn of the twenty-third century, we will have the computer technology on hand to meet the information-transfer challenge of the transporter.

I say this, of course, without any idea of how. It is clear that in order to be able to store in excess of 10^{25} kilobytes of information in any human-scale device, each and every atom of the device will have to be exploited as a memory site. The emerging notions of biological computers, in which molecular dynamics mimics digital logical processes and the 10^{25} or so particles in a macroscopic sample all act simultaneously, seem to me to be the most promising in this regard.

I should also issue one warning. I am not a computer scientist. My cautious optimism may therefore merely be a reflection of my ignorance. However, I take some comfort in the example of the human brain, which is light-years ahead of any existing computational system in complexity and comprehensiveness. If natural selection can develop such a fine information storage and retrieval device, I believe that there is still a long way we can go.

That Quantum Stuff. For some additional cold water of reality moderated by some future hopes, two words: quantum mechanics. At the microscopic level necessary to scan and re-create matter in the transporter, the laws of physics are governed by the strange and exotic laws of quantum mechanics, whereby particles can behave like waves and waves can behave like particles. I am not going to give a course in quantum mechanics here. However, the bottom line is as follows: on microscopic scales, that which is being observed and that which is doing the observation cannot be separated. To make a measurement is to alter a system, usually forever. This simple law can be parameterized in many different ways, but is probably most famous in the form of the Heisenberg uncertainty principle. This fundamental law—which appears to do away with the classical notion of determinism in physics, although in fact at a fundamental level it doesn't—divides the physical world into two sets of observable quantities: the yin and the yang, if you like. It tells us that *no matter what technology is invented in the future,* it is impossible to measure certain combinations of observables with arbitrarily high accuracy. On microscopic scales, one might measure the position of a particle arbitrarily well. However, Heisenberg tells us that we then cannot know its velocity (and hence precisely where it will be in the next instant) very well at all. Or, we might ascertain the energy state of an atom with arbitrary precision. Yet in this case we cannot determine exactly how long it will remain in this state. The list goes on.

These relations are at the heart of quantum mechanics, and they will never go away. As long as we work on scales where the laws of quantum mechanics apply—which, as far as all evidence indicates, is at least larger than the scale at which quantum gravitational effects become significant, or at about 10^{-33} cm—we are stuck with them.

There is a slightly flawed yet very satisfying physical argument that gives some heuristic understanding of the uncertainty principle. Quantum mechanics endows all particles with a wavelike behavior, and waves have one striking property: they are disturbed only when they encounter objects larger than their wavelength (the distance between successive crests). You have only to observe water waves in the ocean to see this behavior explicitly. A pebble protruding from the surface of the water will have no effect on the pattern of the surf pounding the shore. However, a large boulder will leave a region of calm water in its wake.

So, if we want to "illuminate" an atom—that is, bounce light off it so that we can see where it is—we have to shine light of a wavelength small enough so that it will be disturbed by the atom. However, the laws of quantum mechanics tell us that waves of light come in small packets, or quanta, which we call photons (as in starship "photon torpedoes," which in fact are not made of photons). The individual photons of each wavelength have an energy inversely related to their wavelength. The greater the resolution we want, the smaller the wavelength of light we must use. But the smaller the wavelength, the larger the energy of the packets. If we bombard an atom with a high-energy photon in order to observe it, we may ascertain exactly where the atom was when the photon hit it, but the observation process itself—that is, hitting the atom with the photon—will clearly transfer significant energy to the atom, thus changing its speed and direction of motion by some amount.

It is therefore impossible to resolve atoms and their energy configurations with the accuracy necessary to re-create exactly a human pattern. Residual uncertainty in some of the observables is inevitable. What this would mean for the accuracy of the final product after transport is a detailed biological question I can only speculate upon.

This problem was not lost on the Star Trek writers, who were aware of the inevitable constraints of quantum mechanics on the transporter. Possessing something physicists can't usually call upon—namely, artistic license—they introduced "Heisenberg compensators," which allow "quantum resolution" of objects. When an interviewer asked the Star Trek technical consultant Michael Okuda how Heisenberg compensators worked, he merely replied, "Very well, thank you!"

Heisenberg compensators perform another useful plot function. One may wonder, as I have, why the transporter is not also a replicator of life-forms. After all, a replicator exists aboard starships that allows glasses of water or wine to magically appear in each crew member's quarters on voice command. Well, it seems that replicator technology can operate only at "molecular-level resolution" and not "quantum resolution." This is supposed to explain why replication of living beings is not possible. It may also explain why the crew continually complains that the replicator food is never quite the same as the real thing, and why Riker, among others, prefers to cook omelets and other delicacies the old-fashioned way.

Now, shortly after the first edition of this book came out, I got a call from a radio station in Austria, at about 5:00 in the morning, asking me to comment on the new phenomenon called "quantum teleportation." It seemed that an experiment had just been done in Vienna that validated a phenomenon that had first been proposed by scientists at IBM. In fact, as a result, IBM placed a full-page advertisement in *Rolling Stone Magazine,* among others, with one woman calling another on the phone, saying something like, "You can't follow the recipe? Don't worry, I'll teleport the Goulash over." This was supposed to announce an amazing new technique that IBM was working on that they claimed could revolutionize communication.

I expect that were it not for the transporter on Star Trek, the news announcement would probably have been buried on the back page of newspapers, but instead, I was besieged by requests for interviews. Would we soon be transporting like the crew members on Star Trek?

The experiment in Vienna, which has since been refined and extended over the past decade, involved destroying the quantum configuration of a single photon, a particle of light in one location, and instantaneously recreating an identical configuration some distance away. Nothing could sound more like the transporter on Star Trek! Since that experiment, similar experiments have been performed with single atoms, and with molecules containing several atoms. Moreover, the instantaneous transport took place over distances of more than several miles!

As miraculous as this sounds, it is precisely the laws of quantum mechanics, the same laws that make a standard transporter of the type I have described impossible, that are exploited for quantum teleportation. Quantum mechanics tells us if we very carefully set up the initial quantum mechanical configurations, two particles can be "entangled" together in a single quantum mechanical state. Then, if they are separated at a great distance, as long as nothing has interacted with the particles during this time, if one performs an experiment on one particle, such as observing it to be in a specific configuration, the configuration of the other particle will be instantaneously determined. This remains true even if the second particle is now light years away! While this may seem to violate Einstein's rule that information cannot be propagated faster than the speed of light, it turns out that one can show that quantum mechanical entanglement cannot be used to instantaneously send messages, so that cause and effect are preserved.

In any case, quantum entanglement does play a crucial role in performing quantum teleportation. By first entangling the configuration of one particle with another particle, which then interacts with a third particle some distance away, and then performing a specific measurement on the first particle (which changes the state it was in), one can arrange things so that the configuration of the third particle, which is now entangled with the second, would be instantaneously transformed to a state identical to the initial configuration of the first particle before the measurement.

If this sounds a bit confusing, don't worry, because as remarkable as this phenomenon is, I would argue that it wouldn't ever help transport a human, much less Hungarian goulash. The operation of quantum teleportation requires very carefully prepared initial quantum states and then a system that is isolated from its environment throughout the process. Nothing could be further from the situation we exist in, however. We are not quantum objects. If we were, the laws of quantum mechanics would not seem so strange. Macroscopic objects like humans are complex configurations of many particles interacting so frequently with each other and their environment that all quantum mechanical correlations and entanglements are quickly destroyed.

Nevertheless, while quantum teleportation may not ever provide a viable mechanism for moving humans around from place to place, it does have potential uses, including possibly allowing information to be securely communicated from place to place without worry about interception and decryption by third parties. It may also play a role in helping produce a viable quantum computer. During the intermediate steps of the many simultaneous calculations that might be performed by a quantum computer (the reason why a quantum computer might allow calculations to be performed that would take a normal computer longer than the age of the universe to perform), information must be moved around from place to place in the computer without destroying the quantum configurations involved in the calculation, as would be the case if classical measurements were performed. If the configurations were "teleported," however, then the quantum states would be preserved during the computation.

Seeing Is Believing. One last challenge to transporting—as if one more were needed. Beaming down is hard enough. But beaming up may be even more difficult. In order to transport a crew member back to the ship, the sensors aboard the *Enterprise* have to be able to spot the crew member on the planet below. More than that, they need to scan the individual prior to dematerialization and matter-stream transport. So the

Enterprise must have a telescope powerful enough to resolve objects on and often under a planet's surface at atomic resolution. In fact, we are told that normal operating range for the transporter is approximately 40,000 kilometers, or about three times the Earth's diameter. This is the number we shall use for the following estimate.

Everyone has seen photographs of the domes of the world's great telescopes, like the Keck telescope in Hawaii or the Mt. Palomar telescope in California. Have you ever wondered why bigger and bigger telescopes are designed? (It is not just an obsession with bigness—as some people, including many members of Congress, like to accuse science of.) Just as larger accelerators are needed if we wish to probe the structure of matter on ever smaller scales, larger telescopes are needed if we want to resolve celestial objects that are fainter and farther away. The reasoning is simple: Because of the wave nature of light, anytime it passes through an opening it tends to diffract, or spread out a little bit. When the light from a distant point source goes through the telescopic lens, the image will be spread out somewhat, so that instead of seeing a point source, you will see a small, blurred disk of light. Now, if two point sources are closer together across the line of sight than the size of their respective disks, it will be impossible to resolve them as separate objects, since their disks will overlap in the observed image. Astronomers call such disks "seeing disks." The bigger the lens, the smaller the seeing disk. Thus, to resolve smaller and smaller objects, telescopes must have bigger and bigger lenses.

There is another criterion for resolving small objects with a telescope. The wavelength of light, or whatever radiation you use as a probe, must be smaller than the size of the object you are trying to scan, according to the argument I gave earlier. Thus, if you want to resolve matter on an atomic scale, which is about several billionths of a centimeter, you must use radiation that has a wavelength of less than about one-billionth of a centimeter. If you select electromagnetic radiation, this will require the use of either X rays or gamma rays. Here a problem arises right away, because such radiation is harmful to life, and therefore the atmosphere of any Class M planet will filter it out, as our own atmosphere does. The transporter will therefore have to use nonelectromagnetic probes, like neutrinos or gravitons. These have their own problems, but enough is enough . . .

In any case, one can perform a calculation, given that the *Enterprise* is using radiation with a wavelength of less than a billionth of a centimeter and scanning an object 40,000 kilometers away with atomic-scale resolution. I find that in order to do this, the ship would need a telescope with a lens greater than approximately 50,000 kilometers in diameter! Were it any smaller, there would be no possible way even in principle to resolve single atoms. I think it is fair to say that while the *Enterprise-D* is one large mother, it is not that large.

As promised, thinking about transporters has led us into quantum mechanics, particle physics, computer science, Einstein's mass-energy relation, and even the existence of the human soul. We should therefore not be too disheartened by the apparent impossibility of building a device to perform the necessary functions. Or, to put it less negatively, building a transporter would require us to heat up matter to a temperature a million times the temperature at the center of the Sun, expend more energy in a single machine than all of humanity presently uses, build telescopes larger than the size of the Earth, improve present computers by a factor of 1000 billion billion, and avoid the laws of quantum mechanics. It's no wonder that Lieutenant Barclay was terrified of beaming! I think even Gene Roddenberry, if faced with this challenge in real life, would probably choose instead to budget for a landable starship.

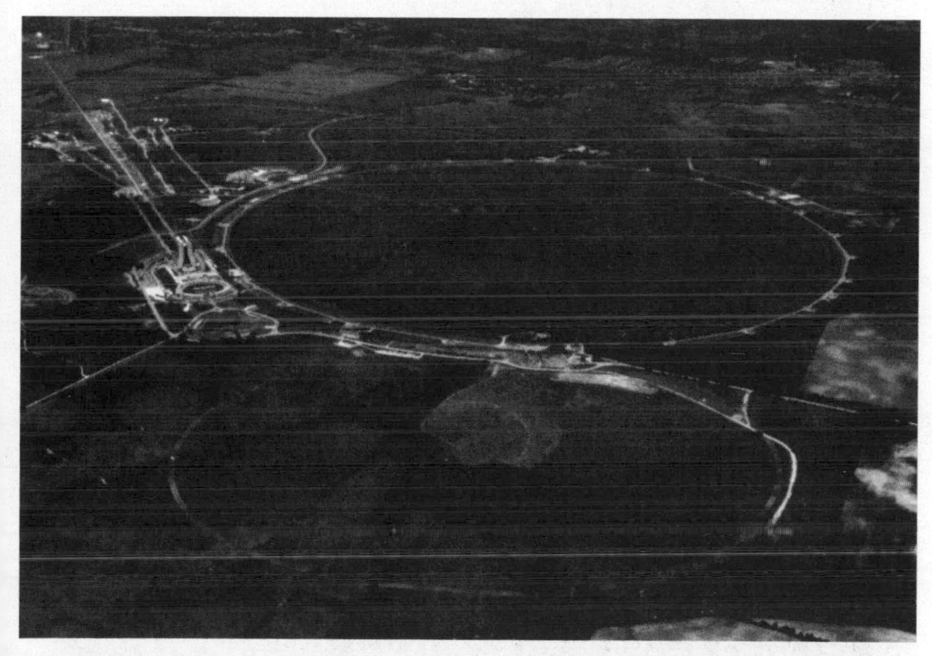

An aerial view of the Fermi National Accelerator laboratory (Fermilab) in Batavia, Illinois, housing the highest energy accelerator in the world, the Tevatron, and the world's largest production and storage facility of antiprotons. The ring housing the 4-mile in circumference accelerator is clearly discernable. The circle in the foreground outlines an accelerator upgrade, the Main Injector, under construction. *(Fermilab Photo)*

John Peoples, director of Fermilab, shown with the antiproton source, which he designed. The antiprotons produced by collisions of protons on a lithium target are stored in a circular beam using the array of magnets shown in the photograph. *(Fermilab Photo)*

A portion of the accelerator tunnel, 4 miles long, located 20 feet below the ground, housing the proton-antiproton beams, and the array of superconducting magnets (lower ring) used to steer and accelerate them to energies approaching 1012 electron volts. *(Fermilab Photo)*

One of the two large detectors at Fermilab built to analyze the high-energy collisions of protons and antiprotons. The 5,000-ton detector is moved in and out of the beam on large rollers. *(Fermilab Photo)*

The Harvard radio-telescope located at Harvard, Massachusetts, used to obtain the data for the Megachannel Extra Terrestrial Array (META) experiment designed to search for the signals of extraterrestrial life in our galaxy.

The META supercomputer array designed to scan millions of channels at a single time in the search for a signal of intelligent life elsewhere in the galaxy.

The Billionchannel Extra Terrestrial Array (BETA) supercomputer that will be a part of the next generation search for extraterrestrial intelligence.

The Andromeda Galaxy (M31). This is the nearest large spiral galaxy similar to our own, located about 2 million light years away. *(Lick Observatory Photograph/Image)*

A photograph of our own galaxy obtained using radio and microwave detectors aboard the Cosmic Background Explorer (COBE) satellite. This is the first true photograph of the Milky Way showing its spiral structure, as edge on from the vantage point of the earth. (NASA/COBE)

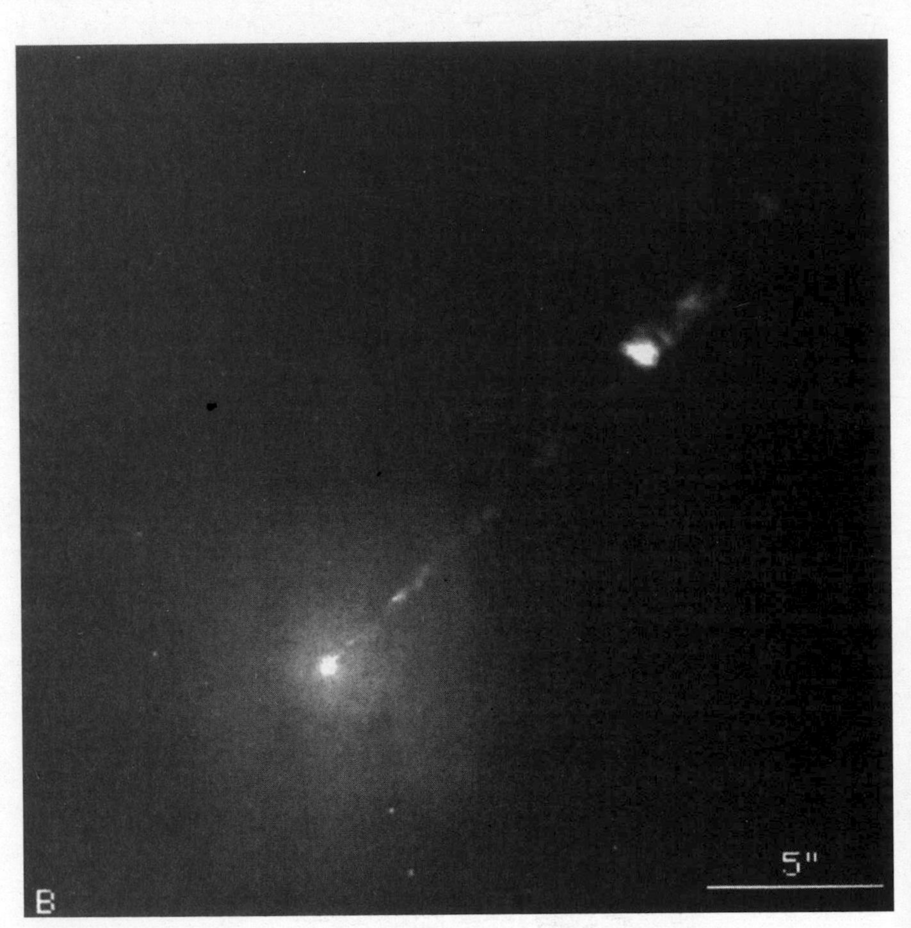

A high resolution photograph of the core of the galaxy M87, which is thought to house a black hole in excess of 2 billion solar masses. The small disk of ionized gas at the very center, almost perpendicular to the large radio jet seen to be emerging from the center is rotating at about 750 kilometers per second, which gives strong dynamical evidence for the existence of such a black hole. *(Holland Ford and NASA)*

6

The Most Bang for Your Buck

Nothing Unreal Exists.

—*Kir-kin-tha's First Law of Metaphysics*
(Star Trek IV: The Voyage Home)

If you are driving west on Interstate 88 out of Chicago, by the time you are 30 miles out of town, near Aurora, the hectic urban sprawl gives way to the gentle Midwestern prairie, which stretches forward and flat as far as you can see. Located slightly north of the interstate at this point is a ring of land marked by what looks like a circular moat. Inside the property, you may see buffalo grazing and many species of ducks and geese in a series of ponds.

Twenty feet below the surface, it is a far cry from the calm pastoral atmosphere above ground. Four hundred thousand times a second, an intense beam of antiprotons strikes a beam of protons head on, producing a shower of hundreds or thousands of secondary particles: electrons, positrons, pions, and more.

This is the Fermi National Accelerator Laboratory, or Fermilab for short. It contains the world's highest-energy particle accelerator, soon to be supplanted by the Large Hadron Collider being built in Geneva, Switzerland, to come online in 2007. But more germane for our purposes is the fact that it is also the world's largest repository of antiprotons. Here, antimatter is not the stuff of science fiction. It is the bread and butter of the thousands of research scientists who use the Fermilab facilities.

It is in this sense that Fermilab and the U.S.S. *Enterprise* bear a certain kinship. Antimatter is crucial to the functioning of a starship: it powers the warp drive. As I mentioned earlier, there is no more efficient way to power a propulsion system (though the warp drive is not, in fact, based on rocket propulsion). Antimatter and matter, when they come into contact, can completely annihilate and produce pure radiation, which travels out at the speed of light.

Obviously, great pains must be taken to make sure that antimatter is "contained" whenever it is stored in bulk. When antimatter containment systems fail aboard starships, as when the *Enterprise*'s system failed after its collision with the *Bozeman,* or when the containment system aboard the *Yamato* failed due to the Iconian computer weapon, total destruction inevitably follows soon afterward. In fact, antimatter containment would be so fundamental to starship operation that it is hard to understand why Federation Lieutenant Commander Deanna Troi was ignorant of the implications of containment loss when she temporarily took over command of the *Enterprise* in the *Next Generation* episode "Disaster," after the ship collided with two "quantum filaments." The fact that she was formally trained only as a psychologist should have been no excuse!

The antimatter containment system aboard starships is plausible, and in fact uses the same principle that allows Fermilab to store antiprotons for long periods. Antiprotons and antielectrons (called positrons) are electrically charged particles. In the presence of a magnetic field, charged particles will move in circular orbits. Thus, if the particles are accelerated in electric fields, and then a magnetic field of appropriate strength is applied, the antiparticles will travel in circles of prescribed sizes. In this way, for example, they can travel around inside a doughnut-shaped container without ever touching the walls. This principle is also used in so called Tokomak devices to contain the high-temperature plasmas in studies of controlled nuclear fusion.

The Antiproton Source for the Fermilab collider contains a large ring of magnets. Once antiprotons are produced, in medium-energy collisions, they are steered into this ring, where they can be stored until they are needed for the highest-energy collisions, which take place in the Tevatron—the Fermilab high-energy collider. The Tevatron is a much larger ring, about four miles in circumference. Protons are injected into the ring and accelerated in one direction, and antiprotons are accelerated in the other. If the magnetic field is carefully adjusted, these two beams of particles can be kept apart throughout most of the tunnel. At specified points, however, the two beams converge and the collisions are studied.

Besides containment, another problem faces us immediately if we want to use a matter-antimatter drive: where to get the antimatter. As far as we can tell, the universe is made mostly of matter, not antimatter. We can confirm that this is the case by examining the content of high-energy cosmic rays, many of which originate well outside our own galaxy. Some antiparticles should be created during the collisions of high-energy cosmic rays with matter, and if one explores the cosmic-ray signatures over wide energy ranges, the antimatter signal is completely consistent with this phenomenon alone; there is no evidence of a primordial antimatter component.

Another possible sign of antimatter in the universe would be the annihilation signature of antiparticle-particle collision. Wherever the two coexist, one would expect to see the characteristic radiation emitted during the annihilation process. Indeed, this is exactly how the *Enterprise* searched for the Crystalline Entity after it had destroyed a new Federation outpost. Apparently the Entity left behind a trace antiproton trail. By looking for the annihilation radiation, the *Enterprise* trailed the Entity and overtook it before it could attack another planet.

While the Star Trek writers got this idea right, they got the details wrong. Dr. Marr and Data search for a sharp "gamma radiation" spike at "10 keV"—a reference to 10 kilo-electron volts, which is a unit of energy of radiation. Unfortunately, this is the wrong scale of energy for the annihilation of protons and antiprotons, and in fact corresponds to no known annihilation signal. The lightest known particle with mass is the electron. If electrons and positrons annihilate, they produce a sharp spike of gamma radiation at 511 keV, corresponding to the mass of the electron. Protons and antiprotons would produce a sharp spike at an energy corresponding to the rest energy of the proton, or about 1 GeV (Giga-electron volt)—roughly a hundred thousand times the energy searched for by Marr and Data. (Incidentally, 10 keV is in the X-ray band of radiation, not the gamma-ray band, which generally corresponds to radiation in excess of about 100 keV, but this is perhaps too fine a detail to complain about.)

In any case, astronomers and physicists have looked for diffuse background signals near 511 keV and in the GeV range as signals of substantial matter-antimatter conflagrations but have not found such signals. This and the cosmic-ray investigations indicate that even if substantial distributions of antimatter were to exist in the universe, they would not be interspersed with ordinary matter.

As most of us are far more comfortable with matter than antimatter, it may seem quite natural that the universe should be made of the former and not the latter. However, there is nothing natural at all about this. In fact, the origin of the excess of matter over antimatter is one of the most interesting unsolved problems in physics today, and is a subject of intense research at the present time. This excess is very relevant to our existence, and thus to Star Trek's, so it seems appropriate to pause to review the problem here.

When quantum mechanics was first developed, it was applied successfully to atomic physics phenomena; in particular, the behavior of electrons in atoms was wonderfully accounted for. However, it was clear that one of the limitations of this testing ground was that such electrons have velocities that are generally much smaller than the speed of light. How to accommodate the effects of special relativity with quantum mechanics remained an unsolved problem for almost two decades. Part of the reason for the delay was that unlike special relativity, which is quite straightforward in application, quantum mechanics required not just a whole new worldview but a vast array of new mathematical techniques. The best young minds in physics were fully occupied in the first three decades of the last century with exploring this remarkable new picture of the universe.

One of those minds was Paul Adrien Maurice Dirac. Like his successor Stephen Hawking, and later Data, he would one day hold the Lucasian Professorship in Mathematics at Cambridge University. Educated by Lord Rutherford, and later training with Niels Bohr, Dirac was better prepared than most to extend quantum mechanics to the realm of the ultrafast. In 1928, like Einstein before him, he wrote down an equation that would change the world. The Dirac equation correctly describes the relativistic behavior of electrons in fully quantum mechanical terms.

Shortly after writing down this equation, Dirac realized that to retain consistency, the mathematics required another particle of equal but opposite charge to the electron to exist in nature. Of course, such a particle was known already—namely, the proton. However, Dirac's equation suggested that this particle should have the same mass as the electron, whereas the proton is almost two thousand times heavier. This discrepancy between observation and the "naive" interpretation of the mathematics remained a puzzle for four years, until the American physicist Carl Anderson discovered, among the cosmic rays bombarding the Earth, a new particle whose mass was identical to the electron's but whose charge was the opposite—that is, positive. This "antielectron" soon became known as the positron.

Since then, it has become clear that one of the inevitable consequences of the merger of special relativity and quantum mechanics is that all particles in nature must possess antiparticles, whose electric charge (if any) and various other properties should be the opposite of their particle partners. If all particles possess antiparticles, then which particles we call particles and which we call antiparticles is completely arbitrary, as long as no physical process displays any bias for particles over antiparticles. In the classical world of electromagnetism and gravity, no such biased process exists.

Now we are left in a quandary. If particles and antiparticles are on an identical footing, why should the initial conditions of the universe have determined that what we call particles should comprise the dominant form of matter?

Surely a more sensible, or at least a more symmetric, initial condition would be that in the beginning the number of particles and antiparticles would have been identical. In this case, we must explain how the laws of physics, which apparently do not distinguish particles from antiparticles, could somehow contrive to produce more of one type than the other. Either there exists a fundamental quantity in the universe—the ratio of particles to antiparticles—which was fixed at the beginning of time and about which the laws of physics apparently have nothing to say, or we must explain the paradoxical subsequent dynamical creation of more matter than antimatter.

In the 1960s, the famous Soviet scientist and later dissident Andrei Sakharov made a modest proposal. He argued that it was possible, if three conditions were fulfilled in the laws of physics during the early universe, to dynamically generate an asymmetry between matter and antimatter even if there was no asymmetry to start with. At the time this proposal was made, there were no physical theories that satisfied the conditions Sakharov laid down. However, in the years since, particle physics and cosmology have both made great strides. Now we have many theories that can, in principle, explain directly the observed difference in abundance between matter and antimatter in nature. Unfortunately, they all require new physics and new elementary particles in order to work; until nature points us in the right direction, we will not know which of them to choose from. Nevertheless, many physicists, myself included, find great solace in the possibility that we may someday be able to calculate from first principles exactly why the matter fundamental to our existence itself exists.

Now, if we had the correct theory, what number would it need to explain? In the early universe, what would the extra number of protons compared to antiprotons need to have been in order to explain the observed excess of matter in the universe today? We can get a clue to this number by comparing the abundance of protons today to the abundance of photons, the elementary particles that make up light. If the early universe began with an equal number of protons and antiprotons, these would annihilate, producing radiation—that is, photons. Each proton-antiproton annihilation in the early universe would produce, on average, one pair of photons. However, assuming there was a small excess of protons over antiprotons, then not all the protons would be annihilated. By counting the number of protons left over after the annihilations were completed, and comparing this with the number of photons produced by those annihilations (that is, the number of photons in the background radiation left over from the big bang), we can get an idea of the fractional excess of matter over antimatter in the early universe.

We find that there is roughly one proton in the universe today for every 10 billion photons in the cosmic background radiation. This means that the original excess of protons over antiprotons was only about *1 part in 10 billion!* That is, for every 10 billion antiprotons in the early universe, there were

10 billion and 1 protons! Even this minuscule excess (accompanied by a similar excess in neutrons and electrons over their antiparticles) would have been sufficient to have produced all the observed matter in the universe—the stars, galaxies, planets—and all that we have come to know and love.

That is how we think the universe got to be made of matter and not antimatter. Aside from its intrinsic interest, the moral of this story for Star Trek is that if you want to make a matter-antimatter drive, you cannot harvest the antimatter out in space, because there isn't very much. You will probably have to make it.

To find out how to do this, we return to the buffalo roaming on the Midwestern plain above the Fermilab accelerator. When thinking about the logistics of this problem, I decided to contact the former director of Fermilab, John Peoples, Jr., who led the effort to design and build its Antiproton Source, and ask if he could help me determine how many antiprotons one could produce and store per dollar in today's dollars. He graciously agreed to help by having several of his staff provide me with the necessary information to make reasonable estimates.

Fermilab produces antiprotons in medium-energy collisions of protons with a lithium target. Every now and then these collisions will produce an antiproton, which is then directed into the storage ring beneath the buffalo. When operating at average efficiency, Fermilab can produce about 50 billion antiprotons an hour in this way. Assuming that the Antiproton Source is operating about 75 percent of the time throughout the year, this is about 6000 hours of operation per year, so Fermilab produces about 300,000 billion antiprotons in an average year.

The cost of those components of the Fermilab accelerator that relate directly to producing antiprotons is about $500 million, in 1995 dollars. Amortizing this over an assumed useful lifetime of 25 years gives $20 million per year. The operating cost for personnel (engineers, scientists, staff) and machinery is about $8 million a year. Next, there is the cost of the tremendous amount of electricity necessary to produce the particle beams and to store the antiprotons. At current Illinois rates, this costs about $5 million a year. Finally, related administrative costs are about $15 million a year. The total comes to some $48 million a year to produce the 300,000 billion antiprotons that Fermilab annually uses to explore the fundamental structure of matter in the universe. This works out to about 6 million antiprotons for a dollar!

Now, this cost is probably higher than it would need to be. Fermilab produces a high-energy beam of antiprotons, and if we required only the antiprotons and not such high energies we might cut the cost, perhaps by a factor of about 2 to 4. So, to be generous, let's assume that using today's technology, one might be able to get from 10 million to 20 million antiprotons for a buck, wholesale.

The next question is almost too obvious: How much bang for this buck? If we convert entirely the mass of one dollar's worth of antiprotons into energy, we would release approximately 1/1000 of a joule, which is the amount of energy required to heat up about 1/4 of a gram of water by about 1/1000 of a degree Celsius. This is nothing to write home about.

Perhaps a better way to picture the potential capabilities of the Fermilab Antiproton Source as the nucleus of a warp core is to consider the energy that might be generated by utilizing every antiproton produced by the Source in real time. The Antiproton Source can produce 50 billion antiprotons an hour. If all these antiprotons were converted into energy, this would result in a power generation of about 1/1000 of a watt! Put another way, you would need about 100,000 Fermilab Antiproton Sources to power a single lightbulb! Given the total annual cost of $48 million to run the Antiproton Source, it would cost at the present time more than the annual budget of the U.S. government to light up your living room in this way.

The central problem is that as things stand today it requires far more energy to produce an antiproton than you would get out by converting its rest mass back into energy. The energy lost during the production process is probably at least a million times more than the energy stored in the antiproton mass. Some much more effective means would be needed for antimatter production before we could ever think of using matter-antimatter drives to propel us to the stars.

It is also clear that if the *Enterprise* were to make its own antimatter, vast new technologies of scale would be needed—not just for cost reduction, but for space reduction. If accelerator techniques were to be utilized, machines that generate far more energy per meter than those of today would be necessary. I might add that this is currently a subject of intense research here on early-twenty-first-century Earth. If particle accelerators, which are our only tools for directly exploring the fundamental structure of matter, are not to become too costly for even international consortiums to build, new technologies for accelerating elementary particles must be developed. (We have already seen that our own government has decided that it is too expensive to build a next-generation accelerator in this country, so a European group will be building one in Geneva, designed to come on line in 2007.) Past trends in the efficiency of energy generation per meter of accelerator suggest that a tenfold improvement may be possible every decade or two. So perhaps in several centuries it will not be unreasonable to imagine a starship-size, antimatter-producing accelerator. Given the current reluctance of governments to support expensive fundamental research at this scale, one might not be so optimistic, but in two centuries a lot of political changes can occur.

Even if one were to make antimatter on board ship, however, one would still have to deal with the fact that to produce each antiproton would invariably use up much more energy than one would get out afterward. Why would one want to expend this energy on antimatter production, when one might turn it directly into propulsion?

The Star Trek writers, always on the ball, considered this problem. Their answer was simple. While energy available in other forms could be used for impulse propulsion and hence sublight speeds, *only* matter-antimatter reactions could be used to power the warp drive. And because warp drive could remove a ship from danger much more effectively than impulse drive, the extra energy expended to produce antimatter might be well worth it in a pinch. The writers also sidestepped the accelerator-based antimatter-production problems by inventing a new method of antimatter production. They proposed hypothetical "quantum charge reversal devices," which would simply flip the charge of elementary particles, so that one could start with protons and neutrons and end up with antiprotons and antineutrons. According to the *Next Generation Technical Manual,* while this process is incredibly power-intensive, there is a net energy loss of only 24 percent—orders of magnitude less than the losses described above for accelerator use.

While all this is very attractive, unfortunately simply flipping the electric charge of a proton is not enough. Consider, for example, that both neutrons and antineutrons are neutral. Antiparticles have all the opposite "quantum numbers" (labels describing their properties) of their matter partners. Since the quarks that make up protons possess many labels other than electric charge, one would have to have many other "quantum reversal devices" to complete the transition from matter to antimatter.

In any case, we are told in the technical manual that, except for emergency antimatter production aboard starships, all Starfleet antimatter is produced at Starfleet fueling facilities. Here antiprotons and antineutrons are combined to form the nuclei of anti-heavy hydrogen. What is particularly amusing is that the Starfleet engineers then add antielectrons (positrons) to these electrically charged nuclei to make neutral anti-heavy-hydrogen atoms—probably because neutral antiatoms sound easier to handle than electrically charged antinuclei to the Star Trek writers. (In fact, antiatoms have recently been produced in the laboratory by a Harvard group working in Geneva to create anti-hydrogen.) Unfortunately, this raises severe containment problems, since magnetic fields, which are absolutely essential for handling substantial amounts of antimatter without catastrophe, work *only* for electrically charged objects! Ah well, back to the drawing board. . . .

The total antimatter fuel capacity of a starship is approximately 3000 cubic meters, stored in various storage pods (on Deck 42 in the *Enterprise-D*). This is claimed to be sufficient for a 3-year mission. Just for fun, let's estimate how much energy one could get out of this much antimatter if it were stored as anti-heavy-hydrogen nuclei. I will assume that the nuclei are transported as a rarefied plasma, which would probably be easier to contain magnetically than a liquid or solid. In this case, 3000 cubic meters could correspond to about 5 million grams of material. If 1 gram per second were consumed in annihilation reactions, this would produce a power equivalent to the total power expended on a daily basis by the human race at the present time. As I indicated earlier in discussing warp drive, one must be prepared to produce at least this much power aboard a starship. One could continue using the fuel at this rate for 5 million seconds, or about 2 months. Assuming that a starship utilizes the matter-antimatter drive for 5 percent of the time during its missions, one might then get the required 3 years' running time out of this amount of material.

Also of some relevance to the amount of antimatter required for energy production is another fact (one that the Star Trek writers have chosen to forget from time to time): matter-antimatter annihilation is an all-or-nothing proposition. It is not continuously tunable. As you change the ratio of matter to antimatter in the warp drive, you will not change the absolute power-generation rate. The relative power versus fuel used will decrease only if some fuel is wasted—that is, if some particles of matter fail to find antimatter to annihilate with, or if they merely collide without annihilating. In a number of episodes ("The Naked Time," "Galaxy's Child," "Skin of Evil") the matter-antimatter ratio is varied, and in the Star Trek technical manual this ratio is said to vary continuously from 25:1 to 1:1 as a function of warp speed, with the 1:1 ratio being used at warp 8 or higher. For speeds higher than warp 8, the amount of reactants is increased, with the ratio remaining unchanged. Changing the amount of reactants and not the ratio should be the proper procedure throughout, as even Starfleet cadets should know. Wesley Crusher made this clear when he pointed out, in the episode "Coming of Age," that the Starfleet exam question on matter-antimatter ratios was a trick question and that there was only one possible ratio—namely, 1:1.

Finally, the Star Trek writers added one more crucial component to the matter-antimatter drive. I refer to the famous dilithium crystals (coincidentally invented by the Star Trek writers long before the Fermilab engineers decided upon a lithium target in their Antiproton Source). It would be unthinkable not to mention them, since they are a centerpiece of the warp drive and as such figure prominently in the economics of the Federation and in various plot developments. (For example, without the economic importance of dilithium, the *Enterprise* would never have been sent to the Halkan system to secure its mining rights, and we would never have been treated to the "mirror universe," in which the Federation is an evil empire!)

What do these remarkable figments of the Star Trek writers' imaginations do? These crystals (known also by their longer formula—2(5)6 dilithium 2(:)1 diallosilicate 1:9:1 heptoferranide) can regulate the matter-antimatter annihilation rate, because they are claimed to be the only form of matter known which is "porous" to antimatter.

I liberally interpret this as follows: Crystals are atoms regularly arrayed in a lattice; I assume therefore that the antihydrogen atoms are threaded through the lattices of the dilithium crystals and therefore remain a fixed distance both from atoms of normal matter and one another. In this way, dilithium could regulate the antimatter density, and thus the matter-antimatter reaction rate.

The reason I am bothering to invent this hypothetical explanation of the utility of a hypothetical material is that once again, I claim, the Star Trek writers were ahead of their time. A similar argument, at least in spirit, was proposed many years after Star Trek introduced dilithium-mediated matter-antimatter annihilation, in order to justify an equally exotic process: cold fusion. During the cold-fusion heyday, which lasted about 6 months, it was claimed that by putting various elements together chemically one could somehow induce the nuclei of the atoms to react much more quickly than they might otherwise and thus produce the same fusion reactions at room temperature that the Sun requires great densities and temperatures in excess of a million degrees to generate.

One of the many implausibilities of the cold-fusion arguments which made physicists suspicious is that chemical reactions and atomic binding take place on scales of the order of the atomic size, which is a factor of 10,000 larger than the size of the nuclei of atoms. It is difficult to believe that reactions taking place on scales so much larger than nuclear dimensions could affect nuclear reaction rates. Nevertheless, until it was realized that the announced results were irreproducible by other groups, a great many people spent a great deal of time trying to figure out how such a miracle might be possible.

Since the Star Trek writers, unlike the cold-fusion advocates, never claimed to be writing anything other than science fiction, I suppose we should be willing to give them a little extra slack. After all, dilithium-mediated reactions merely aid what is undoubtedly the most compellingly realistic aspect of starship technology: the matter-antimatter drives. And I might add that crystals—tungsten in this case, not dilithium—are indeed used to moderate, or slow down, beams of anti-electrons (positrons) in modern-day experiments; here the antielectrons scatter off the electric field in the crystal and lose energy.

There is no way in the universe to get more bang for your buck than to take a particle and annihilate it with its antiparticle to produce pure radiation energy. It is the ultimate rocket-propulsion technology, and will surely be used if ever we carry rockets to their logical extremes. The fact that it may take quite a few bucks to do it is a problem the twenty-third-century politicians can worry about.

7

Holodecks and Holograms

"Oh, we are us, sir. They are also us. So, indeed, we are both us."

—Data to Picard and Riker, in "We'll Always Have Paris"

When Humphrey Bogart said to Ingrid Bergman at the Casablanca airport, "We'll always have Paris," he meant, of course, the memory of Paris. When Picard said something similar to Jenice Manheim at the holodeck re-creation of the Café des Artistes, he may have intended it more literally. Thanks to the holodeck, memories can be relived, favorite places revisited, and lost loves rediscovered—almost.

The holodeck is one of the most fascinating pieces of technology aboard the Enterprise. To anyone already familiar with the nascent world of virtual reality, either through video games or the more sophisticated modern high-speed computers, the possibilities offered by the holodeck are particularly enticing. Who wouldn't want to enter completely into his or her own fantasy world at a moment's notice?

It is so seductive, in fact, that I have little doubt that it would be far more addictive than it is made out to be in the series. We get some inkling of "holodeck addiction" (or "holodiction") in the episodes "Hollow Pursuits" and "Galaxy's Child." In the former, everyone's favorite neurotic officer, Lieutenant Reginald Barclay, becomes addicted to his fantasy vision of the senior officers aboard the *Enterprise,* and would rather interact with them on the holodeck than anywhere else on the ship. In the latter, when Geordi LaForge, who has begun a relationship with a holodeck representation of Dr. Leah Brahms, the designer of the ship's engines, meets the real Dr. Brahms, things become complicated.

Given the rather cerebral pastimes the crew generally engage in on the holodeck, one may imagine that the hormonal instincts driving twentieth-century humanity have evolved somewhat by the twenty-third century (although if this is the case, Will Riker is not representative of his peers). Based on what I know of the world of today, I would have expected that sex would almost completely drive the holodeck. (Indeed, the holodeck would give safe sex a whole new meaning.) I am not being facetious here. The holodeck represents what is so enticing about fantasy, particularly sexual fantasy: actions without consequences, pleasure without pain, and situations that can be repeated and refined at will.

The possible hidden pleasures of the holodeck are merely alluded to from time to time in the series. For example, after Geordi has barged in rather rudely on Reg's private holodeck fantasy, he admits, "I've spent a few hours on the holodeck myself. Now, as far as I'm concerned, what you do on the holodeck is your own business, as long as it doesn't interfere with your work." If that doesn't sound like a twentieth-century admonition against letting the pleasures of the flesh get the better of one, I don't know what does.

I have little doubt that our century's tentative explorations of virtual reality are leading us in the direction of something very much like the holodeck, at least in spirit. Perhaps my concerns will appear as quaint in the twenty-third century as the warning cries that accompanied the invention of television a half century ago. After all, though cries continue because of the surfeit of televised sex and violence, without television there would be no Star Trek.

The danger that we will become a nation of couch potatoes would not apply in a world full of personal holodecks, or perhaps holodecks down at the mall; engaging in holodeck play is far from passive. However, I still find the prospect of virtual reality worrisome, precisely because though it appears real, it is much less scary than real life. The attraction of a world of direct sensual experience without consequences could be overwhelming.

Nevertheless, every new technology has bad as well as good sides and will force adjustments in our behavior. It's probably clear from the tone of this book that I believe technology has on the whole made our lives better rather than worse. The challenge of adjusting to it is just one part of the challenge of being part of an evolving human society.

Be that as it may, the holodeck differs in one striking way from most of the virtual-reality technologies currently under development. At present, through the use of devices that you strap on and that influence your vision and sensory input, virtual reality is designed to put the "scene" inside you. The holodeck takes a more inventive tack: it puts you inside the scene. It does this in part by inventive use of holography and in part by replication.

The principles on which holography is based were first elucidated in 1947, well before the technology was available to fully exploit it, by the British physicist Dennis Gabor, who subsequently won the Nobel Prize for his work. By now, most people are familiar with the use of three-dimensional holographic images on credit cards, and even on the covers of books. The word "hologram" derives from the Greek words for "whole" and "to write." Unlike normal photographs, which merely record two-dimensional representations of three-dimensional reality, holograms give you the whole picture. In fact, it is possible with holography to re-create a three-dimensional image that you can walk around and view from all sides, as if it were the original object. The only way to tell the difference is to try touching it. Only then will you find that there is nothing there to touch.

How can a two-dimensional piece of film, which is what stores the holographic image, record the full information of a three-dimensional image? To answer this we have to think a little about exactly what it is we see when we see something, and what a photograph actually records.

We see objects either because they emit or reflect light, which then arrives at our eyes. When a three-dimensional object is illuminated, it scatters light in many different directions because of this three-dimensionality. If we could somehow reproduce the exact pattern of divergent light created when light is scattered by the actual object, then our eyes would not be able to distinguish the difference between the actual object and the divergent-light pattern *sans* object. By moving our head, for example, we would be able to see features that were previously obscured, because the entire pattern of scattered light from all parts of the object would have been re-created.

How can we first store and then later re-create all this information? We can gain some insight into this question by thinking about what a normal photograph—which stores and later re-creates a two-dimensional image—actually records. When we take a picture, we expose a light-sensitive material to the incoming light, which arrives through the lens of the camera. This light-sensitive material, when exposed to various chemicals, will darken in proportion to the intensity of the light that impinged upon it. (I am discussing black-and-white film here, but the extension to color film is simple—one just coats the film with three different substances, each of which is sensitive to a different primary color of light.)

So, the total information content recorded on a photographic film is the intensity of light arriving at each point on the film. When we develop the film, those points on it that were exposed to a greater intensity of light will react with the development chemicals to become darker, while those not so exposed will remain lighter. The resulting image on the film is a "negative" two-dimensional projection of the original light field. We can project light through this negative onto a light-sensitive sheet of paper to create the final photograph. When we look at it, light hitting the lighter areas of the photograph will be predominantly reflected, while light hitting the darker areas will be absorbed. Thus, looking at the light reflected from the photograph produces a two-dimensional intensity pattern on our retinas, which then allows us to interpret this pattern.

The question then becomes, what more is there to record than just the intensity of light at each point? Once again, we rely on the fact that light is a wave. Because of this fact, more than just intensity is needed to characterize its configuration. Consider the light wave shown below:

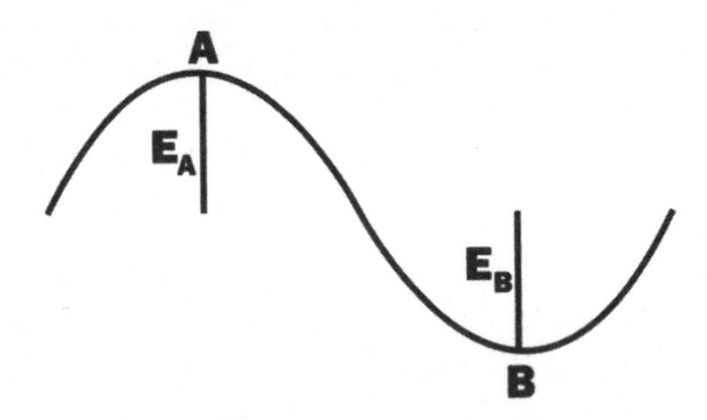

At position A, the wave, which in this case represents the strength of the electric field, has its maximum value, corresponding to an electric field with strength E_A, pointing upward. At point B, the field is exactly the same strength but is pointing downward. Now, if you are sensitive only to the intensity of the light wave, you will find that the field has the same intensity at A as it does at B. However, as you can see, position B represents a different part of the wave from position A. This "position" along the wave is called the *phase.* It turns out that you can specify all the information associated with a wave at a given point by giving its intensity and its phase. So, to record all the information about the light waves scattered by a three-dimensional object, you have to find a way of recording on a piece of film both the intensity and the phase of the scattered light.

This is simple to do. If you split a light beam into two parts and shine one part directly onto the film and let the other part scatter off the object before illuminating the film, then either one of two things can happen. If the two light waves are "in phase"—that is, both have crests coinciding at some point A—then the amplitude of the resulting wave at A will be twice the amplitude of either individual wave, as shown in the figure below:

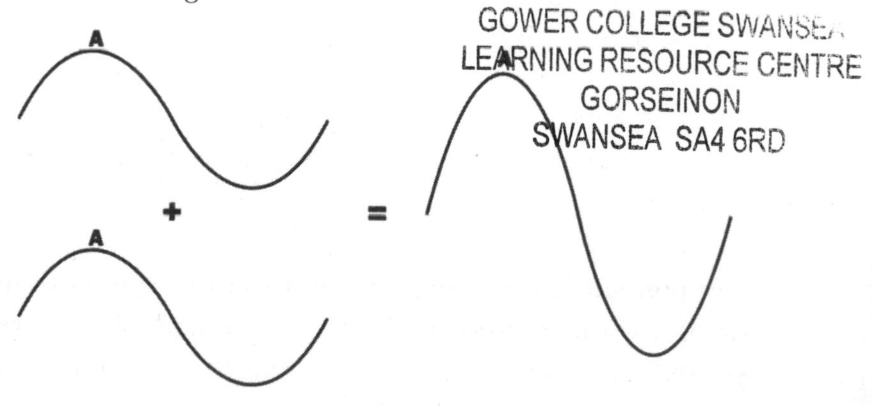

On the other hand, if the two waves are out of phase at point A, then they will cancel each other out, and the resulting "wave" at A will have zero amplitude:

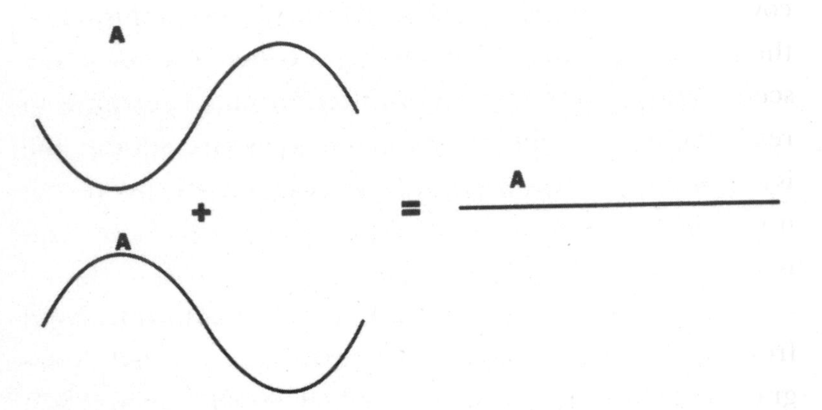

So now, if the film at point A is photographic film, which records intensity only, the pattern recorded will be the "interference pattern" of the two waves—the reference beam and the beam of light scattered by the object. This pattern contains not only the information about the intensity of the scattered light from the object, but information about its phases as well. If one is clever, one can extract this information to re-create a three-dimensional image of the object that scattered the light.

In fact, it turns out that one doesn't have to be all that clever. If one merely illuminates this photographic film with a source of light of the same wavelength as the original light that produced the interference pattern, an image of the object will be created exactly where the object was in relation to the film, when you look through the film. If you move your head to one side, you will be able to "look around" the edges of the re-created object. If you cover up most of the piece of film and hold it closer to your eyes and look through the uncovered part, you will still see the entire object! In this sense, the experience is just like looking through a window at a scene outdoors, except that the scene you are seeing isn't really there. The light coming to your eyes through the film is affected in just such a way as to make your eyes believe that it has been scattered off objects, which you then "see." This is a hologram.

Normally, in order for the reference light and the light from the scattered object to be carefully controlled, holograms are made using laser light, which is coherent and well collimated. However, so-called "white light" holograms exist, which can be illuminated by ordinary light to produce the same effect.

One can be trickier and arrange, just as one can using various lenses, for the image of the objects you see to appear to be between you and the film, and you will see before you the three-dimensional image of an object, which you can walk around and view from all sides. Or you can arrange for the light source to be in front of the film instead of behind it—as in the holograms on credit cards.

Presumably the former sort of hologram is used on the holodeck, and to re-create the image of a doctor in the sick bay, as in the *Voyager* series. What's more, in order to create such holograms, one would not need to use the original objects to make the holographic images. Digital computers are now sophisticated enough to do "ray tracing"—that is, they can calculate the pattern of light scattered from any hypothetical object you want to draw on the screen, and illuminate it from any angle. In the same way, the computer could determine the configuration of the interference pattern that would be caused by merging the light from a direct beam with the scattered light from an object. This computer-generated interference pattern could be projected onto a transparent screen, and when this screen is illuminated from behind, a three-dimensional image is produced of an object that in fact never existed. If the computer is fast enough, it can project a continuously changing interference pattern on the screen, thereby producing a moving three-dimensional image. So the holographic aspect of the holodeck is not particularly far-fetched.

However, holograms aren't all there is to the holodeck. As noted, they have no corporeal integrity. You can walk through one—or shoot through one, as was evidenced by the wonderful holographic representations created by Spock and Data to trick the Romulans in the episode "Unification." This incorporeality simply will not do for the objects one would like to interact with—that is, touch—on the holodeck. Here techniques that are more esoteric are required, and the Star Trek writers have turned to the transporter, or at least to the replicators, which are less sophisticated versions of the transporter. Presumably, using transporter technology, matter is replicated and moved around on the holodeck to resemble exactly the beings in question, in careful coordination with computer programs that control the voices and movements of the re-created beings. Similarly, the replicators reproduce the inanimate objects in the scene—tables, chairs, and so forth. This "holodeck matter" owes its form to the pattern held in the replicator buffer. When the transporter is turned off or the object is removed from the holodeck, the matter can then disassemble as easily as it would if the pattern buffer were turned off during the beaming process. Thus, creatures created from holodeck matter can be trapped on the holodeck, as the fictional detectives Cyrus Redblock and Felix Leach found to their dismay in the *Next Generation* episode "The Big Goodbye," and as Sherlock Holmes's nemesis Professor Moriarty surmised and then attempted to overcome in several other episodes.

So here is how I envisage the holodeck: holograms would be effective around the walls, to give one the impression of being in a three-dimensional environment that extended to the horizon, and the transporter-based replicators would then create the moving "solid" objects within the scene. Since holography is realistic, while (as I have explained earlier) transporters are not, one would have to find some other way of molding and moving matter around in order to make a workable holodeck. Still, one out of two technologies in hand isn't bad.

Indeed, I recently visited a modern three-dimensional video "cave" at the Banff School of Fine Art which was a close approximation to the realistic part of the "holodeck" I envisaged above. While holograms were not used to create the three-dimensional images, one could nevertheless manipulate computer generated images on the walls of the cave which, when viewed with 3D glasses, moved around the room under one's control.

In any case, where does all this leave the pure holograms, like the holographic doctor of the *Voyager* series? The answer is, Absolutely nowhere. With just the scattered light and no matter around, I'm afraid that these images would not be very effective at lifting, manipulating, or probing. However, a good bedside manner and compassionate words of advice, which are at the heart of good medical practice, can be dispensed by a hologram as easily as by the real thing.

SECTION THREE

THE INVISIBLE UNIVERSE, OR THINGS THAT GO BUMP IN THE NIGHT

In which we speak of things that may exist but are not yet seen—extraterrestrial life, multiple dimensions, and an exotic zoo of other physics possibilities and impossibilities

8

The Search for Spock

"It's difficult to work in a group when you are omnipotent."

—Q, upon joining the crew of the *Enterprise*, in "Déjà Q"

"Restless aggression, territorial conquest, and genocidal annihilation . . . whenever possible. . . . The colony is integrated as though it were in fact one organism ruled by a genome that constrains behavior as it also enables it. . . . The physical superorganism acts to adjust the demographic mix so as to optimize its energy economy. . . . The austere rules allow of no play, no art, no empathy."

The Borg are among the most frightening, and intriguing, species of alien creature ever portrayed on the television screen. What makes them so fascinating, from my point of view, is that some organism like them seems plausible on the basis of natural selection. Indeed, although the paragraph quoted above provides an apt description of the Borg, it is not taken from a Star Trek episode. Rather it appears in a review of Bert Hölldobler and Edward O. Wilson's *Journey to the Ants,* and it is a description not of the Borg but of our own terrestrial insect friends.[1] Ants have been remarkably successful on an evolutionary scale, and it is not hard to see why. Is it impossible to imagine a cognizant society developing into a similar communal superorganism? Would intellectual refinements such as empathy be necessary to such a society? Or would they be a hindrance?

Gene Roddenberry said that the real purpose of the starship *Enterprise* was to serve as a vehicle not for space travel but for storytelling. Beyond all the technical wizardry, even a techie such as myself recognizes that what makes Star Trek tick is drama, the same grand themes that have driven storytelling since the Greek epics—love, hate, betrayal, jealousy, trust, joy, fear, wonder. . . . We all connect most closely with stories that illuminate those human emotions that govern our own lives. If warp drive were used merely to propel unmanned probes, if the transporters were developed merely to move soil samples, if medical scanners were utilized merely on plant life, Star Trek would never have made it past the first season.

Indeed, the "continuing mission" of the starship *Enterprise* is not to further explore the laws of physics but "to explore strange new worlds, to seek out new life and new civilizations." What makes Star Trek so fascinating—and so long-lived, I suspect—is that this allows the human drama to be extended far beyond the human realm. We get to imagine how alien species might develop to deal with the same problems and issues that confront humanity. We are exposed to new imaginary cultures, new threats. It provides some of the same fascination as visiting a foreign country for the first time does, or as one sometimes gets from reading history and discovering both what is completely different and what is exactly the same about the behavior of people living centuries apart.

We must, of course, suspend disbelief for such entertainment. Remarkably, almost all alien species encountered by the *Enterprise* are humanlike, and they all speak English! (In their defense, the Star Trek writers invented, in the sixth season of *The Next Generation,* a rationale for this. The archeologist Richard Galen apparently discovers that a wide variety of these civilizations share genetic material, which was seeded in the primordial oceans of many different worlds by some very ancient civilization. This is a notion reminiscent of the Nobel laureate Francis Crick's [only partly] tongue-in-cheek theory of Panspermia.)[2] This has not escaped the notice of any trekker, and it was perhaps most colorfully put to me by the theoretical physicist and Nobel laureate Sheldon Glashow, who said of the aliens, "They all look like people with elephantiasis!" Nevertheless, he is willing to ignore, as are most trekkers, these plot contrivances in order to appreciate the Star Trek writers' exploration of alien psychologies. Hollywood screenwriters are generally neither scientists nor engineers, and thus it is natural to expect that most of their creative energy would go into designing alien cultures rather than alien biology.

And creative they have been. Besides the Borg and the omnipotent prankster Q, over two hundred specific life-forms populated the Star Trek universe at the point when I gave up counting. Our galaxy is apparently full of other intelligent civilizations, some more advanced and some less advanced. Some—like the Federation, the Klingons, the Romulans, and the Cardassians—control large empires, while others exist in isolation on single planets or in the emptiness of space.

The discovery of extraterrestrial intelligence could be, as emphasized by the practitioners of the ongoing search, the greatest discovery in the history of the human race. Certainly it is hard to imagine a discovery that might change our view of ourselves and our place in the universe more than this. Nevertheless, after three decades of concerted searching, we have yet to find any definitive evidence for any form of life outside our own planet. One might find this surprising. Certainly, if there is life out there, it seems inevitable that we should find it, just as many of the civilizations that independently emerged on several continents here on Earth eventually ran into each other, sometimes traumatically.

Nevertheless, when one thinks in some detail about the likelihood of discovering intelligent life elsewhere in the universe, the daunting nature of the search becomes clear. Consider, for example, that some other civilization in the galaxy was informed somehow of exactly where to look among the 400 billion or so stars in the Milky Way to find a planet that could support life. Say further that they were directed to look in the direction of our Sun. What is the probability even then that they would discover our existence? Life has existed on Earth for much of the 4.5 billion years since it formed. Yet only in the past half century or so have we been transmitting any signals of our existence. Furthermore, only in the past 25 years or so have we had radiotelescopes sufficiently powerful to serve as radio beacons for observation by other civilizations. Thus, in the 4.5 billion years during which aliens might have been scanning the Earth from space, they could have discovered us only during the last half century. Assuming that an alien civilization chose to make its observations at some random time during the planet's history, the possibility of discovering our existence would be about 1 in 100 million. And I remind you, this applies only if they knew exactly where to look!

There have been whole books written about the possibility of life existing elsewhere in the galaxy, and also about the possibility of detecting it. Estimates for the number of advanced civilizations range from millions on the high side to one on the low side (liberally interpreting our own civilization as advanced). It is not my purpose to review all the arguments in depth here. I would like, however, to describe some of the more interesting physical arguments related to the origin of the sorts of life the *Enterprise* was sent out to discover, and to discuss some of the strategies currently being employed here on Earth to search for it.

The a priori argument that life should exist elsewhere in our galaxy seems to me to be compelling. As noted, there are roughly 400 billion stars in our galaxy. It would seem truly remarkable if our Sun were the only one around which intelligent life developed. One can propose what on the surface seems like a more sophisticated argument to estimate the probability that life like ourselves occurs elsewhere, starting with obvious questions such as: "What is the probability that most stars have planets?" or "What is the probability that this [particular] star will live long enough to sustain life on a planetary system?" and then moving on to planetary matters, such as "Is this planet big enough to hold an atmosphere?" or "What is the likelihood of its having undergone sufficient early volcanism to produce enough water on the surface?" or "What is the probability of its having a moon either massive enough or close enough to produce tides sufficient to make tidal pools where life might originate, but not daily tidal waves?" While I will discuss some of these issues, the problem with trying to determine realistic probabilities is, first, that many of the relevant parameters are undetermined and, second, that we do not know how all the parameters are correlated. It is difficult enough to determine accurately the probability of everyday events. When one sets out to estimate a sequence of very small probabilities, the operational significance of such an attempt often becomes marginal.

One should also remember that even if one derives a well-defined probability, its interpretation can be pretty subtle. For example, the probability of any specific sequence of events—such as the fact that I am sitting in this specific type of chair typing at this specific computer (among all the millions of computers manufactured each year), in this specific place (among all the possible cities in the world), at this specific time of day (among the 86,400 seconds in each day) is vanishingly small. The same can be said for any other set of circumstances in my life. Likewise, in the inanimate world, the probability that, say, a radioactive nucleus will decay at the exact moment it does is also vanishingly small. However, we do not calculate such probabilities. We ask, rather, how likely it is that the nucleus will decay in some nonzero time interval, or how much more probable a decay is at one time compared to another time.

When one is attempting to estimate the probabilities of life in the galaxy, one has to be very careful not to overrestrict the sequence of events one considers. If one does, and people have, one is likely to conclude that the probability that life formed on Earth when it did is infinitesimally small, which is sometimes used as an argument for the existence of Divine intervention. However, as I have just indicated, the same vanishingly small probability could be assigned to the likelihood that the stoplight I can see out my window will turn red while I am waiting in my car there at precisely 11:57 A.M. on June 3, 2008. This does not mean, however, that such a thing won't happen.

The important fact to recognize is that *life did form* in the galaxy at least once. I cannot overemphasize how important this is. Based on all our experience in science, nature rarely produces a phenomenon just once. We are a test case. The fact that we exist proves that the formation of life is possible. Once we know that life can originate here in the galaxy, the likelihood of it occurring elsewhere is vastly increased. (Of course, as some evolutionary biologists have argued, it need not develop an intelligence.)

⋐⋑

While our imaginations are no doubt far too feeble to consider all the combinations of conditions which might give rise to intelligent life, we can use our own existence to ask what properties of the universe were essential or important in our own evolution.

We first begin with the universe as a whole. I have already mentioned one cosmic coincidence: that there was one extra proton produced in the early universe for every 10 billion or so protons and antiprotons. Without these extra little guys, matter would have annihilated with antimatter, and there would be no matter left in the universe today, intelligent or otherwise.

The next obvious feature of the universe in which we live is that it is old, very old. It took intelligent life about 4.5 billion years to develop on Earth. Hence, our existence requires a universe that accommodated our arrival by lasting billions of years. The current best estimate for the age of our universe is between about 14 billion years, which is plenty long enough. It turns out, however, that it is not so easy a priori to design a universe that expands, as our universe does, without either recollapsing very quickly in a reverse of the big bang—a big crunch—or expanding so fast that there would have been no time for matter to clump together into stars and galaxies. The initial conditions of the universe, or some dynamical physical process early in its history, would have to be very finely tuned to get things just right.

This has become known as the "flatness" problem, and understanding it has become one of the central issues in cosmology. Gravitational attraction due to the presence of matter tends to slow the expansion of the universe. As a result, two possibilities appeared to remain. Either there was enough matter in the universe to cause the expansion to halt and reverse (a "closed" universe), or there was not (an "open" universe). What was surprising about the universe is that when we added up all the matter we estimated out there, the amount we found was close to the borderline between these two possibilities—a "flat" universe, in which the observed expansion would slow but never quite stop in any finite amount of time.

What makes this particularly surprising is that as the universe evolves, if it is not exactly flat then it deviates more and more from being flat as time goes on. Since the universe is probably at least 10 billion years old today, and observations suggest that the universe is close to being flat today, then at much earlier times it must have been immeasurably close to being flat. It is hard to imagine how this could happen at random without some physical process enforcing it. Some 25 years ago, a candidate physical process was invented. Known as "inflation," it is a ubiquitous process that can occur due to quantum mechanical effects in the early universe.

Recall that empty space is not really empty but that quantum fluctuations in the vacuum can carry energy. It turns out that it is possible, as the nature of forces between elementary particles evolves with temperature in the early universe, for the energy stored as quantum fluctuations in empty space to be the dominant form of energy in the universe. This vacuum energy can repel gravitationally rather than attract. It is hypothesized that the universe went through a brief inflationary phase, during which it was dominated by such vacuum energy, resulting in a very rapid expansion. One can show that when this period ends and the vacuum energy is transferred into the energy of matter and radiation, the universe can easily end up being flat to very high precision.

However, another, perhaps more severe, problem remains. In fact Einstein first introduced the problem when he tried to apply his new general theory of relativity to the universe. At that time, it was not yet known that the universe was expanding; rather, the universe was believed to be static and unchanging on large scales. So Einstein had to figure out some way to stop all this matter from collapsing due to its own gravitational attraction. He added a term to his equations called the cosmological constant, which essentially introduced a cosmic repulsion to balance the gravitational attraction of matter on large scales. Once it was recognized that the universe is not static, Einstein realized that there was no need for such a term, whose addition he called "the biggest blunder" he had ever made.

Unfortunately, as in trying to put the toothpaste back into the tube, once the possibility of a cosmological constant is raised, there is no going back. If such a term is possible in Einstein's equations then we must explain why it is absent in the observed universe. In fact, the vacuum energy I described above produces exactly the effect that Einstein sought to produce with the cosmological constant. So the question becomes, How come such vacuum energy is not overwhelmingly dominant in the universe today?—or, How come the universe isn't still inflating at a large rate?

This question was one of the most profound unanswered questions in physics. Every calculation we perform with the theories we now have suggests that the vacuum energy should be many orders of magnitude larger today than it is allowed to be on the basis of our observations. To make matters even more confusing, when the first edition of this book appeared I mentioned that it had begun to look like the energy of empty space, while not overwhelmingly large, isn't zero after all, but is still large enough to affect the present and future evolution of the universe. Around that time, by analyzing cosmological data then available, including the age of the universe, the density of matter, and large scale structure estimates, I, along with a colleague from Chicago, in fact proposed that the data would be consistent with a flat universe only if as much as 70 percent of the energy of the universe existed as energy associated with empty space not associated with galaxies and clusters.

This heretical proposal was strikingly confirmed in 1998 by observations of the expansion rate of the universe. Recall that vacuum energy, like a cosmological constant, produces a repulsive force. By measuring a specific type of exploding star or supernova in distant galaxies in order to determine the distance to these objects, while at the same time measuring the speed of recession of these galaxies, observers discovered, much to their surprise, that the expansion of the universe appears to be speeding up. This is completely counterintuitive in a universe in which gravity is normally attractive and should be slowing the expansions. Nevertheless, the observed acceleration could be explained if 70 percent of the energy of the universe resided in empty space and 30 percent resided in matter.

At the same time, independent measurements of the curvature of space on large scales demonstrated that to high precision the universe is flat, as was theoretically expected. Meanwhile, determinations of the total mass associated with galaxies and clusters have now definitively demonstrated that about only 30 percent of the total mass required to produce a flat universe exists in these systems. The only way these two observations could be consistent is in fact if 70 percent of the total energy of the universe must exist in some other form. Everything now points to the fact that the dominant energy in the universe exists in the form of something that looks very much like a cosmological constant!

While the near flatness of the universe may have been necessary for the eventual formation of life on Earth and elsewhere, the inferred value of the cosmological constant suggests an even greater possible fine tuning problem. It was expected in advance that some theoretical trick would explain why the cosmological constant must be zero today. But now apparently the energy of empty space is not zero. So why is it so small? Interestingly, it turns out that if it were perhaps even an order of magnitude or two bigger, then one can show that galaxies would never have formed in the early universe—the repulsive force introduced by this "dark energy" in empty space would have more than compensated for the attraction of gravity between the nascent pre-galactic lumps. If there had been no galaxies, there would have been no stars, or planets, no aliens, and no Star Trek writers.

A universe without Star Trek writers may be too horrible to imagine, but nevertheless, scientists have done so. In fact at a fundamental microphysical level, there is also a whole slew of cosmic coincidences that allowed life to form on Earth. If any one of a number of fundamental physical quantities in nature was slightly different, then the conditions essential for the evolution of life on Earth would not have existed. For example, if the very small mass difference between a neutron and proton (about 1 part in 1000) were changed by only a factor of 2, the abundance of elements in the universe, some of which are essential to life on Earth, would be radically different from what we observe today. Along the same lines, if the energy level of one of the excited states of the nucleus of the carbon atom were slightly different, then the reactions that produce carbon in the interiors of stars would not occur and there would be no carbon—the basis of organic molecules—in the universe today.

It is hard to know how much emphasis to put on these coincidences. It is not surprising, since we *have* evolved in this universe, to find that the constants of nature happen to have the values that allowed us to evolve in the first place. Nevertheless, the incredibly unexpected apparent value of the cosmological constant has caused a growing number of theorists to suspect that perhaps the present value of the cosmological constant is also telling us something. Somehow, for example, that our observed universe is part of a meta-universe, or multiverse, containing many possible different universes, in each of which the fundamental constants of nature could be different. In those universes which are incompatible with the evolution of life, no one would be around to measure anything. To paraphrase the argument of the Russian cosmologist Andrei Linde, who happens to subscribe to this form of what is known as the "anthropic principle," it is like an intelligent fish wondering why the universe in which it lives (the inside of a fish bowl) is made of water. The answer is simple: if it weren't made of water, the fish wouldn't be there to ask the question.

Moreover, it turns out that there are several theoretical proposals that suggest our universe might actually be part of a multiverse. Inflation, for example, with its early phase of rapid expansion, suggests our observable universe might be only a small part of a much bigger whole. If inflation were to end different ways in different parts of this region, only in those parts in which the laws of physics were appropriate would life form.

More recently another theoretical possibility has been proposed which is even more exotic. String Theory, of which I shall have more to say shortly, requires a host of new, thus far undetected dimensions, and with these, a host of new possibilities for four-dimensional universes. In fact, one of the many drawbacks of this theory has been that even if it were true, it apparently cannot predict why our universe looks the way it does. With the anthropic principle, that defect has been claimed as an advantage. String theorists now talk of a string "landscape," with potentially 10^{500} different possible four-dimensional universes, each of which might have different laws of physics, and have argued that our universe may look the way it does on anthropic, not fundamental grounds.

I want to stress that if these ideas are true, the physics of our universe would not be due to anything truly fundamental, but merely an environmental accident, associated with the fact that we happen to be living in it. As such, many physicists, who became physicists because they were motivated to discover the fundamental laws governing the universe, have not been thrilled with this possibility.

These issues, while interesting, may not be empirically resolvable for some time, and are perhaps thus best left to philosophers or science fiction writers. Let us then accept the fact that the universe *has* managed to evolve, both microscopically and macroscopically, in a way that is conducive to the evolution of life. We next turn to our own home, the Milky Way galaxy.

When we consider which systems in our own galaxy may house intelligent life, the physics issues are much more clear-cut. Given that there exist stars in the Milky Way which, from all estimates, are at least 10 billion years old, while life on Earth is no older than about 3.5 billion years, we are prompted to ask how long life could have existed in our galaxy before it arose on Earth.

When our galaxy began to condense out of the universal expansion some 10–12 billion years ago, its first generation stars were made up completely of hydrogen and helium, which were the only elements created with any significant abundance during the big bang. Nuclear fusion inside these stars continued to convert hydrogen to helium, and once this hydrogen fuel was exhausted, helium began to "burn" to form yet heavier elements. These fusion reactions will continue to power a star until its core is primarily iron. Iron cannot be made to fuse to form heavier elements, and thus the nuclear fuel of a star is exhausted. The rate at which a star burns its nuclear fuel depends on its mass. Our own Sun, after 5 billion years of burning hydrogen, is not even halfway through the first phase of its stellar evolution. Stars of 10 solar masses—that is, 10 times heavier than the Sun—burn fuel at about 1000 times the rate the Sun does. Such stars will go through their hydrogen fuel in less than 100 million years, instead of in the Sun's 10-billion-year lifetime.

What happens to one of these massive stars when it exhausts its nuclear fuel? Within seconds of burning the last bit, the outer parts of the star are blown off in an explosion known as a supernova, one of the most brilliant fireworks displays in the universe. Supernovae briefly shine with the brightness of a billion stars. At the present time, they are occurring at the rate of about two or three every 100 years in the galaxy. Almost 1000 years ago, Chinese astronomers observed a new star visible in the daytime sky, which they called a "guest star." This supernova created what we now observe telescopically as the Crab Nebula. It is interesting that nowhere in Western Europe was this transient object recorded. Church dogma at the time declared the heavens to be eternal and unchanging, and it was much easier not to take notice than to be burned at the stake. Almost 500 years later, European astronomers had broken free enough of this dogma so that the Danish astronomer Tycho Brahe was able to record the next observable supernova in the galaxy.

Many of the heavy elements created during the stellar processing, and others created during the explosion itself, are dispersed into the interstellar medium, and some of this "stardust" is incorporated in gas that collapses to form another star somewhere else. Over billions of years, later generations of stars—so-called Population 1 stars, like our Sun—form, and any number of these can be surrounded by a swirling disk of gas and dust, which would coalesce to form planets containing heavy elements like calcium, carbon, and iron. Out of this stuff we are made. Every atom in our bodies was created billions of years ago, in the fiery furnace of some long dead star, or stars. I find this one of the most fascinating and poetic facts about the universe and have written a whole book, titled *Atom*, about it: we are all literally star children.

Now, it would not be much use if a planet like the Earth happened to form near a very massive star. As we have seen, such stars evolve and die within the course of 100 million years or so. Only stars of the mass of our Sun or less will last longer than 5 billion years in a stable phase of hydrogen burning. It is hard to imagine how life could form on a planet orbiting a star that changed in luminosity by huge amounts over the course of such evolution. Conversely, if a star smaller and dimmer than our Sun should have a planetary system, any planet warm enough to sustain life would probably be so close in as to be wracked by tidal forces. Thus, if we are going to look for life, it is a good bet to look at stars not too different from our own. As it happens, the Sun is a rather ordinary member of the galaxy. About 25 percent of all stars in the Milky Way—some 100 billion of them—fall in the range required. Most of these are older even than the Sun and could therefore, in principle, have provided sites for life up to 4 billion to 5 billion years before the Sun did.

On to the Earth. What is it about our fair green-blue planet that makes it special? In the first place, it is in the inner part of the solar system. This is important, because the outer planets have a much higher percentage of hydrogen and helium—much closer to that of the Sun. Most of the heavy elements in the disk of gas and dust surrounding the Sun at its birth appear to have remained in the inner part of the system. Thus, one might expect potential sites for life to be located at distances smaller than, say, the distance of Mars from a 1-solar-mass star.

Next, as Goldilocks might have said, the Earth is just right—not too big or too small, too cold or too hot. Since the inner planets probably had no atmosphere when they formed, these had to be generated by gases produced by volcanoes. At least some of the water on the Earth's surface was also produced in this fashion. A smaller planet might well have radiated heat from its surface rapidly enough to prevent a large amount of volcanism. Presumably this is the case with Mercury and the Moon. Mars is a borderline case, while Earth and Venus have successfully developed an atmosphere. Recent measurements of radioactive gas isotopes in the terrestrial rocks suggest that after an initial period of bombardment, in which the Earth was created by the accretion of infalling material over a period of 100 million to 150 million years about 4.5 billion years ago, volcanism produced about 85 percent of the atmosphere within a few million years. So, again, it is not surprising that organic life formed on Earth rather than on other planets in the solar system, and one might expect similar proclivities elsewhere in the galaxy—on Class M planets, as they are called in the Star Trek universe.

The next question is how quickly life, followed by intelligent life, might take to evolve, based on our experience with the Earth. The answer to the first part of the question is: remarkably quickly. Fossil relics of blue-green algae about 3.5 billion years old have been discovered, and various researchers have argued that life was already flourishing as long as 4 billion years ago. Within a few 100 million years of the earliest possible time that life could have evolved on Earth, it did. This is very encouraging.

Of course, from the time life first began on Earth until complex multicellular structures, and later intelligent life, evolved, almost 3 billion years may have elapsed. There is every reason to believe that this time was governed more by physics than biology. In the first place, the Earth's original atmosphere contained no oxygen. Carbon dioxide, nitrogen, and trace amounts of methane, ammonia, sulfur dioxide, and hydrochloric acid were all present, but not oxygen. Not only is oxygen essential for the advanced organic life-forms on Earth, it plays another important role. Only when there is sufficient oxygen in the atmosphere can ozone form. Ozone, as we are becoming more and more aware, is essential to life on Earth because it screens out ultraviolet radiation, which is harmful to most life-forms. It is therefore not surprising that the rapid explosion of life on Earth began only after oxygen was abundant.

Recent measurements indicate that oxygen began building up in the atmosphere about 2 billion years ago and reached current levels within 600 million years after that.

While oxygen had been produced earlier, by photosynthesis in the blue-green algae of the primordial oceans, it could not at first build up in the atmosphere. Oxygen reacts with so many substances, such as iron, that whatever was photosynthetically produced combined with other elements before it could reach the atmosphere. Eventually, enough materials in the ocean were oxidized so that free oxygen could accumulate in the atmosphere. (This process never took place on Venus because the temperature was too high there for oceans to form, and thus the life-forming and life-saving blue-green algae never arose there.)

So, after conditions were really ripe for complex life-forms, it took about a billion years for them to evolve. Of course, it is not clear at all that this is a characteristic timescale. Accidents such as evolutionary wrong turns, climate changes, and cataclysmic events that caused extinctions affected both the biological timescale and the end results.

Nevertheless, these results indicate that intelligent life can evolve in a rather short interval on the cosmic timescale—a billion years or so. The extent of this time-frame has to do with purely physical factors, such as heat production and chemical reaction rates. Our terrestrial experience suggests that even if we limit our expectations of intelligent life to the organic and aerobic—surely a very conservative assumption, and one that the Star Trek writers were willing to abandon (the silicon-based Horta is one of my favorites)—planets surrounding several-billion-year-old stars of about 1 solar mass are good candidates.

Granting that the formation of organic life is a robust and relatively rapid process, what evidence do we have that its fundamental ingredients—namely, organic molecules and other planets—exist elsewhere in the universe? Here, again, recent results lead to substantial optimism. Organic molecules have been observed in asteroids, comets, meteorites, and interstellar space. Some of these are complex molecules, including amino acids, the building blocks of life. Microwave measurements of interstellar gas and dust grains have led to the identification of dozens of organic compounds, some of which are presumed to be complex hydrocarbons. There is little doubt that organic matter is probably spread throughout the galaxy.

Finally, what about planets? It has long been believed that most stars have planets around them. Certainly a fair fraction of observed stars have another stellar companion, in so-called binary systems. Moreover, many young stars are observed to have circumstellar disks of dust and gas, which are presumably the progenitors of planets. Various numerical models for predicting the distribution of planetary masses and orbits in such disks suggest that they will produce on average at least one Earthlike planet at an Earthlike distance from its star. But the proof of the pudding is in the tasting, and over the past decade or so, since the first edition of this book came out, over 100 different planets have been discovered to be orbiting around stars other than our Sun. In some cases these planets exist in the least expected places, planets the size of Jupiter located closer than Mercury is to our Sun, planets orbiting the collapsed core of a star that had exploded, etc. There have even been planets discovered in the so-called "habitable" zone near their stars, i.e., the zone where, like the Earth, liquid water may exist. All of these discoveries confirm that planetary formation is anything but rare.

I do not want to lose the forest for the trees here. It is almost miraculous that the normal laws of physics and chemistry, combined with an expanding universe more than some 10 billion years old, lead to the evolution of conscious minds that can study the universe out of which they were born. Nevertheless, while the circumstances that led to life on Earth are special, they appear to be by no means peculiar to Earth. The arguments above suggest that there could easily be over a billion possible sites for organic life in our galaxy. And since our galaxy is merely one out of 400 billion galaxies in the observable universe, I find it hard to believe that we are alone. Moreover, as I noted earlier, most Population 1 stars were formed earlier than our Sun was—up to 5 billion years earlier. Given the time frame discussed above, it is likely that intelligent life evolved on many sites billions of years before our Sun was even born. In fact, it might be expected that most intelligent life in the galaxy existed before ours. Thus, depending upon how long intelligent civilizations persist, the galaxy could be full of civilizations that have been around literally billions of years longer than we have. On the other hand, given our own history, such civilizations may well have faced the perils of war and famine, and many may not have made it past a few thousand years; in this case, most of the intelligent life in the universe would be long gone. As one researcher cogently put the issue over 20 years ago, "The question of whether there is intelligent life out there depends, in the last analysis, upon how intelligent that life is."[3]

So, how will we ever know? Will we first send out starships to explore strange new worlds and go where no one has gone before? Or will we instead be discovered by our galactic neighbors, who have tuned in to the various Star Trek series as these signals move at the speed of light throughout the galaxy? I think neither will be the case, and I am in good company.

In the first place, we have clearly seen how daunting interstellar space travel would be. Energy expenditures beyond our current wildest dreams would be needed—warp drive or no warp drive. Recall that to power a rocket by propulsion using matter-antimatter engines at something like 3/4 the speed of light for a 10-year round-trip voyage to just the nearest star would require an energy release that could fulfill the entire current power needs in the United States for more than 100,000 years! This is dwarfed by the power that would be required to actually warp space. Moreover, to have a fair chance of finding life, one would probably want to be able to sample at least several thousand stars. I'm afraid that even at the speed of light this couldn't be done anytime in the next millennium.

That's the bad news. The good news, I suppose, is that by the same token we probably don't have to worry too much about being abducted by aliens. They, too, have probably figured out the energy budget and will have discovered that it is easier to learn about us from afar.

So, do we then devote our energies to broadcasting our existence? It would certainly be much cheaper. We could send to the nearest star system a 10-word message, which could be received by radio antennae of reasonable size, for much less than a dollar's worth of electricity. However—and here again I borrow an argument from the Nobel laureate Edward Purcell—if we broadcast rather than listen, we will miss most of the intelligent life-forms. Obviously, those civilizations far ahead of us can do a much better job of transmitting powerful signals than we can. And since we have been in the radio-transmission business for only 80 years or so, there are very few societies less advanced than we are that could still have the technology to receive our signals. So, as my mother used to say, we should listen before we speak. Although as I write this, I suddenly hope that all those more advanced societies aren't thinking exactly the same thing.

But what do we listen to? If we have no idea which channel to turn to in advance, the situation seems hopeless. Here we can be guided by Star Trek. In the *Next Generation* episode

"Galaxy's Child," the *Enterprise* stumbles upon an alien life-form that lives in empty space, feeding on energy. Particularly tasty is radiation with a very specific frequency—1420 million cycles per second, having a wavelength of 21 cm.

In the spirit of Pythagoras, if there were a Music of the Spheres, surely this would be its opening tone. Fourteen hundred and twenty megahertz is the natural frequency of precession of the spin of an electron as it encircles the atomic nucleus of hydrogen, the dominant material in the universe. It is, by a factor of at least 1000, the most prominent radio frequency in the galaxy. Moreover, it falls precisely in the window of frequencies that, like visible light, can be transmitted and received through an atmosphere capable of supporting organic life. And there is very little background noise at this frequency. Radioastronomers have used this frequency to map out the location of hydrogen in the galaxy—which is, of course, synonymous with the location of matter—and have thus determined the galactic shape. Any species intelligent enough to know about radio waves and about the universe will know about this frequency. It is the universal homing beacon. Thirty-six years ago, the astrophysicists Giuseppe Cocconi and Philip Morrison proposed that this is the natural frequency to transmit at or listen to, and no one has argued with this conclusion since.

Hollywood not only guessed the right frequency to listen to but helped put up the money to do the listening. While small-scale listening projects have been carried out for more than 30 years, the first large-scale comprehensive program came on line in the autumn of 1985, when Steven Spielberg threw a big copper switch that formally initiated Project META, which stands for Megachannel Extra Terrestrial Array. The brainchild of electronics wizard Paul Horowitz at Harvard University, META was located at the Harvard/Smithsonian 26-meter radiotelescope in Harvard, Massachusetts, and funded privately by the Planetary Society, including a $100,000 contribution from Mr. ET himself. META used an array of 128 parallel processors to scan simultaneously 8,388,608 frequency channels in the range of 1420 megahertz and its so-called second harmonic, 2840 megahertz. More than 4 years of data were taken, and META covered the sky three times looking for an extraterrestrial signal.

Of course, you have to be clever when listening. First, you have to recognize that even if a signal is sent out at 1420 megahertz, it may not be received at this frequency. This is because of the infamous Doppler effect—a train whistle sounds higher when it is approaching and lower when it is receding. The same is true for all radiation emitted by a moving source. Since most of the stars in the galaxy are moving at velocities of several hundreds of kilometers per second relative to us, you cannot ignore the Doppler shift. (The Star Trek writers haven't ignored it; they added "Doppler compensators" to the transporter to account for the relative motion of the starship and the transporter target.) Reasoning that the transmitters of any signal would have recognized this fact, the META people have looked at the 1420 megahertz signal as it might appear if shifted from one of three reference frames: (a) one moving along with our local set of stars, (b) one moving along with the center of the galaxy, and (c) one moving along with the frame defined by the cosmic microwave background radiation left over from the big bang. Note that this makes it easy to distinguish such signals from terrestrial signals, because terrestrial signals are all emitted in a frame fixed on the Earth's surface, which is not the same as any of these frames. Thus terrestrial signals have a characteristic "chirp" when present in the META data.

What would an extraterrestrial signal involve? Cocconi and Morrison suggested that we might look for the first few prime numbers: 2, 3, 5, 7, 11, 13. . . . In fact, this is precisely the series that Picard taps out in the episode "Allegiance," when he is trying to let his captors know that they are dealing with an intelligent species. Pulses from, say, a surface storm on a star are hardly likely to produce such a series. Indeed, this is precisely the signal that the Jodie Foster character in Carl Sagan and Ann Druyan's *Contact* discovers when she first gets evidence of extraterrestrial intelligence. The META people have searched for an even simpler signal: a uniform constant tone at a fixed frequency. Such a "carrier" wave is easy to search for.

Horowitz and his collaborator, the Cornell astronomer Carl Sagan, reported on an analysis of the 5 years of META data. Thirty-seven candidate events, out of 100,000 billion signals detected, were isolated. However, none of these "signals" has ever repeated. Horowitz and Sagan prefer to interpret the data as providing no definitive signal thus far. As a result, they have been able to put limits on the number of highly advanced civilizations within various distances of our Sun which have been trying to communicate with us.

Since that early experiment, and since the first edition of this book appeared, a host of new, more sophisticated SETI (Search for Extraterrestrial Intelligence) experiments have been performed, and are ongoing. In fact readers of this book can participate in the search via the SETI at home program, which uses unused home computer cycles to help analyze data. Other, new types of optical SETI programs, looking for blinking light signals, have been embarked upon. What was once a lonely research field is now full of excitement.

Nevertheless, in spite of the incredible complexity of the present search efforts, only a small range of frequencies has actually been explored, and the power of the signals capable of being detected is immense, much greater than the total power generated by all human activity on Earth. Yet there is no cause for pessimism. It is a big galaxy.

The search goes on. The fact that we have not yet heard anything should not dissuade us. It is something like what my friend Sidney Coleman, a physics professor at Harvard, once told me about buying a house: You shouldn't get discouraged if you look at a hundred and don't find anything. You only have to like one. . . . A single definitive signal, as improbable as it is that we will ever hear one, would change the way we think about the universe, and would herald the beginning of a new era in the evolution of the human race.

And for those of you who are disheartened at the idea that our first contact with extraterrestrial civilizations will not be made by visiting them in our starships, remember the Cytherians, a very advanced civilization encountered by the *Enterprise* who made outside contact with other civilizations not by traveling through space themselves but by bringing space travelers to them. In some sense, that is exactly what we are doing as we listen to the signals from the stars.

9

The Menagerie of Possibilities

"That is the exploration that awaits you! Not mapping stars and studying nebula, but charting the unknown possibilities of existence."

—Q to Picard, in "All Good Things . . ."

In the course of more than 19 TV-years of the various Star Trek series, the writers have had the opportunity to tap into some of the most exciting ideas from all fields of physics. Sometimes they get it right; sometimes they blow it. Sometimes they just use the words that physicists use, and sometimes they incorporate the ideas associated with them. The topics they have dealt with read like a review of modern physics: special relativity, general relativity, cosmology, particle physics, time travel, space warping, and quantum fluctuations, to name just a few.

In this penultimate chapter, I thought it might be useful to make a brief presentation of some of the more interesting ideas from modern physics which the Star Trek writers have borrowed—in particular, concepts I haven't concentrated on elsewhere in the book. Because of the diversity of the ideas, I give them here in glossary form, with no particular ordering or theme. In the last chapter, I will follow a similar format—this time to sample the most blatant physics blunders in the series, as chosen by myself, selected fellow-physicists, and various trekkers. In both chapters, I have restricted my lists to the top ten examples; there are a lot more to choose from.

The Scale of the Galaxy and the Universe. Our galaxy is the stage on which the Star Trek drama is enacted. Throughout the series, galactic distance scales of various sorts play a crucial role in the action. Units from AUs (for Astronomical Unit: 1 AU is 93 million miles, the distance from the Earth to the Sun), which were used to describe the size of the V'ger cloud in the first Star Trek movie, to light-years are bandied about. In addition, various features of our galaxy are proposed, including a "Great Barrier" at the center (*Star Trek V: The Final Frontier*) and, in the original series, a "galactic barrier" at the edge (cf. "Where No Man Has Gone Before," "By Any Other Name," and "Is There in Truth No Beauty?"). It seems appropriate, therefore, in order to describe the playing field where Star Trek's action takes place, to offer our own present picture of the galaxy and its neighbors, and of distance scales in the universe.

Because of the large number of digits required, one rarely expresses astronomical distances in conventional units such as miles or kilometers. Instead, astronomers have created several fiducial lengths that seem more appropriate. One such unit is the AU, the distance between the Earth and the Sun. This is the characteristic distance scale of the solar system, with the former outermost planet, Pluto, being nearly 40 AU from the Sun. In *Star Trek: The Motion Picture,* the V'ger cloud is described as 82 AU in diameter, which is remarkably big—bigger, in fact, than the size of our solar system!

For comparison with interstellar distances, it is useful to express the Earth-Sun distance in terms of the time it takes light (or the time it would take the *Enterprise* at warp 1) to travel from the Sun to the Earth—about 8 minutes. (This should be the time it would take light to travel to most Class M planets from their suns.) Thus, we can say that an AU is 8 light-minutes. By comparison, the distance to the nearest star, Alpha Centauri—a binary star system where the inventor of warp drive, Zefrem Cochrane, apparently lived—is about 4 light-years! This is a characteristic distance between stars in our region of the galaxy. It would take rockets, at their present rate of speed, more than 10,000 years to travel from here to Alpha Centauri. At warp 9, which is about 1500 times the speed of light, it would take about 6 hours to traverse 1 light-year.

The distance of the Sun from the center of the galaxy is approximately 25,000 light-years. At warp 9, it would take almost 15 years to traverse this distance, so it is unlikely that Sybok, having commandeered the *Enterprise,* would have been able to take her to the galactic center, as he did in *Star*

Trek V: The Final Frontier, unless the *Enterprise* was essentially already there.

The Milky Way is a spiral galaxy, with a large central disk of stars. It is approximately 100,000 light-years across and a few thousand light-years deep. The *Voyager,* tossed 70,000 light-years away from Earth in the first episode of that series, would thus indeed be on the other side of the galaxy. At warp 9, the ship would take about 50 years to return to the neighborhood of our Sun from that distance, which is why *Voyager* ultimately needed to find a shortcut to make its way home.

At the center of our galaxy is a large galactic bulge—a dense conglomeration of stars—several thousand light-years across. It is thought to harbor a black hole of about a million solar masses. Black holes ranging from 100,000 to more than a billion solar masses are likely at the center of many other galaxies.

A roughly spherical halo of very old stars surrounds the galaxy. The conglomerations of thousands of stars called globular clusters found here are thought to be among the oldest objects in our galaxy, perhaps as old as 12–13 billion years according to our current methods of dating—more ancient even than the "black cluster" in the episode "Hero Worship," which was said to be 9 billion years old. An even larger spherical halo, consisting of "dark matter" (about which more later), is thought to encompass the galaxy. This halo is invisible to all types of telescopes; its existence is inferred from the motion of stars and gas in the galaxy, and it may well contain 10 times as much mass as the observable galaxy.

The Milky Way is an average-size spiral galaxy, containing a few hundred billion stars. There are approximately 400 billion galaxies in the observable universe, each containing more or less that many stars! Of the galaxies we see, roughly 70 percent are spiral; the rest are somewhat spherical in shape and are known as elliptical galaxies. The largest of them are giant ellipticals more than 10 times as massive as the Milky Way.

Most galaxies are clustered in groups. In our local group, the nearest galaxies to the Milky Way are small satellite galaxies orbiting our own. These objects, observable in the Southern Hemisphere, are called the Large and Small Magellanic Clouds. It is about 2 million light-years to the nearest large galaxy, the Andromeda galaxy—home to the Kelvans, who attempt to take over the *Enterprise* and return to their home galaxy in the original-series episode "By Any Other Name." At warp 9, the voyage would take approximately 1,000 years!

Because of the time it takes light to travel, as we observe farther and farther out, we are also observing farther and farther back in time. The farthest we can now observe with electromagnetic sensors is back to a time when the universe was about 300,000 years old. Before then, matter existed as a hot ionized gas opaque to electromagnetic radiation. When we look out in all directions, we see the radiation emitted when matter and radiation finally "decoupled." This is known as the cosmic microwave background. Observing it, most recently with the COBE satellite launched by NASA in 1989, we get a picture of what the universe looked like when it was only about 300,000 years old. The picture revealed structures that were probably imprinted in the earliest moments of the big bang and was so significant that two of the principal investigators on COBE were awarded the Nobel Prize in Physics in 2006.

Finally, the universe itself is expanding uniformly. As a result, distant galaxies are observed to be receding from us—and the farther away they are, the faster they are receding, at a rate directly proportional to their distance from us. This rate of expansion, characterized by a quantity called the Hubble constant, is such that galaxies located 10 million light-years from us are moving away at an average rate of about 150 to 300 kilometers per second. Working backward, we find that all the observed galaxies in the universe would converge about 14 billion years ago, at the time of the big bang. As I earlier described, because most of the energy in the universe apparently resides in empty space, the rate of expansion of the universe is speeding up, not slowing down.

Dark Matter. As I mentioned above, our galaxy is apparently immersed in a vast sea of invisible material.[1] By studying the motion of the stars, of hydrogen gas clouds, and even of the Large and Small Magellanic Clouds around the galactic center, and using Newton's laws relating the velocity of orbiting objects to the mass pulling them, it has been determined that there is a roughly spherical halo of dark material stretching out to distances perhaps 10 times as far from the center of the galaxy as we are. This material accounts for at least 90 percent of the mass of the Milky Way. Moreover, as we observe the motion of other galaxies, including the ellipticals, and also the motion of groups of galaxies, we find that there is more matter associated with these systems than we can account for on the basis of the observable material. The clustered mass in the entire observable universe therefore seems to be dominated by dark matter. This material, over 90 percent of which is dark, is currently thought to make up about 30 percent of the total energy of the universe, with the remaining 70 percent residing in the energy of empty space.

The notion of dark matter has crept into both the *Next Generation* and *Deep Space Nine,* as well as the *Voyager* and *Enterprise* series, and in an amusing way. For example, in the *Voyager* episode "Cathexis," the ship enters a "dark matter nebula," which, as you might imagine, is like a dark cloud, so that you cannot see into it. The *Enterprise* had already encountered similar objects, including the "black cluster" mentioned earlier. In *Deep Space Nine,* Benjamin Sisko also seeks refuge in a dark matter nebula. However, the salient fact about dark matter is not that it shields light in any way but that it does not shine—that is, emit radiation—and does not even absorb significant amounts of radiation. If it did either, it would be detectable by telescopes. If you were inside a dark matter cloud, as we probably are, you would not even see it. Indeed, earlier in time, but later in production, in the *Enterprise* series Captain Archer discovers a somewhat more realistic dark matter nebulae, which is indeed invisible. Ultimately he and T'Pol fire a succession of warheads into it, exciting the dark matter and causing it to shine. While not impossible to imagine, for the particles we currently imagine as the best candidates for dark matter, this is a stretch.

The question of the nature, origin, and distribution of dark matter is probably one of the most exciting unresolved issues in cosmology today. Since this unknown material dominates the mass density of the universe, its distribution must have determined how and when the observable matter gravitationally collapsed to create the galactic clusters, galaxies, stars, and planets that make the universe so interesting to us. Our very existence is directly dependent on this material.

Even more interesting are the strong arguments that the dark matter may be made of particles completely different from the protons and neutrons that make up normal matter. Independent limits on the amount of normal matter in the universe, based on calculations of nuclear reaction rates in the early universe and the subsequent formation of light elements, as well as observations of the cosmic microwave background, imply that there are not enough protons and neutrons to account for the dark matter around galaxies and clusters. Moreover, it seems that in order for the small fluctuations in the initial distribution of matter to have collapsed in the hot plasma of the early universe to form the galaxies and clusters we observe today, some new type of elementary particle—of a kind that does not interact with electromagnetic radiation—had to be involved. If the dark matter is indeed made of some new type of elementary particle, then:

(a) the dark matter is not just "out there," it is in this room as you are reading this book, passing imperceptibly through your body. These exotic elementary particles would not clump into astronomical objects; they would form a diffuse "gas" streaming throughout the galaxy. Since they interact at best only very weakly with matter, they would be able to sail through objects as big as the Earth. Indeed, examples of such particles already exist in nature—notably, neutrinos (particles that should be familiar to trekkers, and which I will later discuss).

(b) the dark matter might be detected directly here on Earth, using sophisticated elementary-particle-detection techniques. Various detectors designed with a sensitivity to various dark matter candidates are currently being constructed.

(c) the detections of such particles might revolutionize elementary particle physics. It is quite likely that these objects are remnants of production processes in the very early universe, well before it was 1 second old, and would thus be related to physics at energy scales comparable to or even beyond those we can directly probe using modern accelerators.

Of course, as exciting as this possibility is, we are not yet certain that the dark matter is entirely made from exotic stuff.

There are many ways of putting protons and neutrons together so that they do not shine. For example, if we populated the galaxy with snowballs, or boulders, these would be difficult to detect. Perhaps the most plausible possibility for this scenario is that there are many objects in the galaxy which are almost large enough to be stars but are too small for nuclear reactions to start occurring in their cores. Such objects are known as brown dwarfs, and Data and his colleagues aboard the *Enterprise* have discussed them (for instance, in "Manhunt"). In fact, there are ways to determine whether or not brown dwarfs—known in this context as MACHOs (for Massive Astrophysical Compact Halo Objects)—make up a significant component of the dark matter halo around the Milky Way galaxy. While these objects are not directly observable, if one of them were to pass in front of a star the star's light would be affected by the MACHO's gravity in such a way as to make the star appear brighter. This "gravitational lensing" phenomenon was first predicted by Einstein back in the 1930s, and we now have the technology to detect it. Several experiments in the 1990s observed literally millions of stars in our galaxy to see if this lensing phenomena took place. While some lenses were indeed observed, the data have made it clear that, as expected, MACHOs cannot contribute a significant amount to the dark matter halo surrounding our galaxy.

Neutron Stars. These objects are, as you will recall, all that is left of the collapsed cores of massive stars that have undergone a supernova. Although they typically contain a mass somewhat in excess of the mass of our Sun, they are so compressed that they are about the size of Manhattan! Once again, the Star Trek writers have outdone themselves in the nomenclature department. The *Enterprise* has several times encountered material expelled from a neutron star—a material that the writers have dubbed "neutronium." Since neutron stars are composed almost entirely of neutrons held so tightly together that the star is basically one huge atomic nucleus, the name is a good one. The Doomsday machine in the episode of the same name was apparently made of pure neutronium, which is why it was impervious to Federation weapons. However, in order for this material to be stable it has to be under the incredibly high pressure created by the gravitational attraction of a stellar mass of material only 15 kilometers in radius. In the real world, such material exists only as part of a neutron star.

The *Enterprise* has had several close calls near neutron stars. In the episode "Evolution," when the Nanites began eating the ship's computers, the crew was in the act of studying a neutron star that was apparently about to erupt as it accreted material. In the episode "The Masterpiece Society," the *Enterprise* must deflect a stellar core fragment hurtling toward Moab IV.

There are no doubt millions of neutron stars in the galaxy. Most of these are born with incredibly large magnetic fields inside them. If they are spinning rapidly, they make wonderful radio beacons. Radiation is emitted from each of their poles, and if the magnetic field is tilted with respect to the spin axis, a rotating beacon is created. On Earth, we detect these periodic bursts of radio waves and call their sources pulsars. Rotating out in space, they make the best clocks in the universe. The pulsar signals can keep time to better than one microsecond per year. Moreover, some pulsars produce more than 1000 pulses per second. This means that an object that is essentially a huge atomic nucleus with the mass of the Sun and 10 to 20 kilometers across is rotating over 1000 times each second. Think about that. The rotation speed at the neutron star surface is therefore almost half the speed of light! Pulsars are one illustration of the fact that nature produces objects more remarkable than any the Star Trek writers are likely to invent.

Other Dimensions. As James T. Kirk slowly drifts in and out of this universe in "The Tholian Web," we find that the cause is a "spatial interphase" briefly connecting different dimensional planes, which make up otherwise "parallel universes." Twice before in the series, Kirk encountered parallel universes—one made of antimatter, in "The Alternative Factor," and the other accessed via the transporter, in "Mirror, Mirror." In *The Next Generation,* we have the Q-continuum, Dr. Paul Manheim's nonlinear time "window into other dimensions," and of course subspace itself, containing an infinite number of dimensions, which aliens, like the ones who kidnapped Lieutenant Riker in "Schisms," can hide in.

The notion that somehow the four dimensions of space and time we live in are not all there is has had great tenacity in the popular consciousness. Recently a Harvard psychiatrist wrote a successful book (and apparently got in trouble with the Medical School) in which he reported on his analysis of a variety of patients, all of whom claimed they had been abducted by aliens. In an interview, when asked where the aliens came from and how they got here, he is reported to have suggested, "From another dimension."

This love affair with extra dimensions was no doubt influenced by the special theory of relativity, as I have described in my recent book, *Hiding in the Mirror.* Once three-dimensional space was tied with time to make four-dimensional spacetime by Hermann Minkowski, it was natural to suppose that the process might continue. Moreover, once general relativity demonstrated that what we perceive as the force of gravity can be associated with the curvature of spacetime, it was not outrageous to speculate that perhaps other forces might be associated with curvature in yet other dimensions.

Among the first to speculate on this idea were the German mathematician Theodor Kaluza in 1919 and, independently, the Swedish physicist Oskar Klein in 1926. They proposed that electromagnetism could be unified with gravity in a five-dimensional universe. Perhaps the electromagnetic force is related to some "curvature" in a fifth dimension, just as the gravitational force is due to curvature in four-dimensional spacetime.

This is a very pretty idea, but it has problems. In fact, in any scenario in which one envisages extra dimensions in the universe, one has to explain *why we don't experience these dimensions as we do space and time.* Klein's proposed answer to this question is very important, because it crops up again and again when physicists consider the possibility of higher dimensions in the universe.

Consider a cylinder and an intelligent bug. As long as the circumference of the cylinder is large compared to the size of the bug, then the bug can move along both dimensions and will sense that it is crawling on a two-dimensional surface.

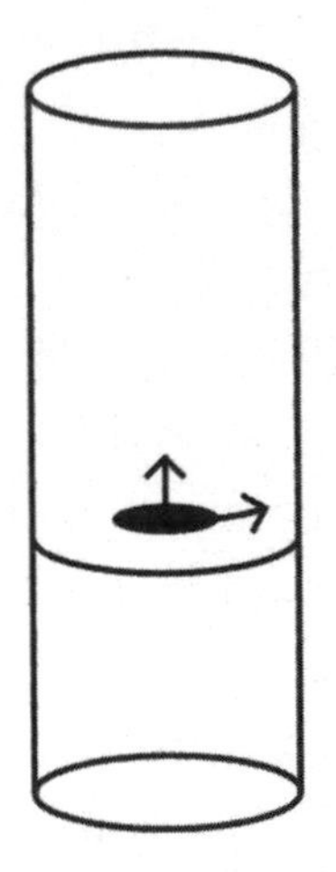

However, if the circumference of the cylinder becomes very small, then as far as the bug is concerned it is crawling on a one-dimensional object—namely, a line or a string—and can move only up or down:

Now think how such a bug might actually find out that there is another dimension, corresponding to the circumference of the cylinder. With a microscope, it might be able to make out the "string's" width. However, the wavelength of radiation needed to resolve sizes this small would have to be on order of the diameter of the cylinder or smaller, because, as I noted in chapter 5, waves scatter off only those objects that are at least comparable to their wavelength. Since the energy of radiation increases as its wavelength decreases, it would require a certain minimum energy of radiation to resolve this "extra dimension."

If somehow a fifth dimension were "curled up" in a tight circle, then unless we focused a lot of energy at a small point, we would not be able to send waves traveling through it to probe its existence, and the world would continue to look to us to be effectively four-dimensional. After all, we know that space is three-dimensional because we can probe it with waves traveling in all three dimensions. If the only waves that can be sent into the fifth dimension have much more energy than we can produce even in high-energy accelerators, then we cannot experience this extra dimension.

In spite of its intrinsic interest, the Kaluza-Klein theory cannot be a complete theory. First, it does not explain *why* the fifth dimension would be curled up into a tiny circle. Second, we now know of the existence of two other fundamental forces in nature beyond electromagnetism and gravity—the strong nuclear force and the weak nuclear force. Why stop at a fifth dimension? Why not include enough extra dimensions to accommodate all the fundamental forces?

In fact, modern particle physics has raised just such a possibility. The modern effort, which took off again shortly after the first edition of this book appeared, is called, as I have already mentioned, string theory. It focused initially on extending the general theory of relativity so that a consistent theory of quantum gravity could be constructed. In the end, however, the goal of a unified theory of all interactions has resurfaced.

I have already noted the challenges faced in developing a theory wherein general relativity is made consistent with quantum mechanics. The key difficulty in this effort is trying to understand how quantum fluctuations in spacetime can be handled. In elementary particle theory, quantum excitations in fields—the electric field, for example—are manifested as elementary particles, or quanta. If one tries to understand quantum excitations in the gravitational field—which, in general relativity, correspond to quantum excitations of spacetime—the mathematics leads to nonsensical predictions.

The advance of string theory was to suppose that at microscopic levels, typical of the very small scales (that is, 10^{-33} cm) where quantum gravitational effects might be important, what we think of as pointlike elementary particles actually could be resolved as vibrating strings. The mass of each particle would correspond in some sense to the energy of vibration of these strings.

The reason for making this otherwise rather outlandish proposal is that it was discovered as early as the 1970s that such a theory requires the existence of particles having the properties that quantum excitations in spacetime—known as gravitons—should have. General relativity is thus in some sense imbedded in the theory in a way that may be consistent with quantum mechanics.

However, a quantum theory of strings cannot be made mathematically consistent in 4 dimensions, or 5, or even 6. It turns out that such theories can exist consistently only in 10 or more dimensions! Indeed, Lieutenant Reginald Barclay, while he momentarily possessed an IQ of 1200 after having been zapped by a Cytherian probe, had quite a debate with Albert Einstein on the holodeck about which of these possibilities was more palatable in order to incorporate quantum mechanics in general relativity.

This plethora of dimensions may seem an embarrassment, but it was quickly recognized that like many embarrassments it also presented an opportunity. Perhaps all the fundamental forces in nature could be incorporated in a theory of 10 or more dimensions, in which all the dimensions but the four we know curl up with diameters on the order of the Planck scale (10^{-33} cm)—as Lieutenant Barclay surmised they must—and are thus unmeasurable today.

Alas, this great hope has remained no more than that. After 35 years, we have no idea whether the tentative proposals of string theory can produce a Theory of Everything. Indeed, the theory has kept morphing with the times, so that strings themselves may not be an important part of the theory, and even the number of dimensions associated with the theory is uncertain, as is any mechanism by which only four dimensions would be left large. And finally, as I earlier alluded, ultimately the theory may not even predict an unambiguous universe like ours, but rather a host of universes, one of which might resemble ours. The theory thus would resemble a "theory of anything," rather than a "theory of everything," or even a "theory of something."

So, the moral of this saga is that Yes, Virginia, there may be extra dimensions in the universe. In fact, there is now some reason to expect them. However, these extra dimensions are not the sort that might house aliens who could then abduct psychiatric patients (or Commander Riker, for that matter). They cannot be mixed up with the four dimensions of spacetime in a way that would allow macroscopic objects to drift from one place to another in space by passing through another dimension, as "subspace" seems to allow in the Star Trek universe.

Nevertheless, in the past decade several interesting new possibilities have been proposed in theoretical physics that more closely resemble the extra dimensions of Star Trek. It was recognized that if only gravity, which is an extremely weak force, could act in extra dimensions, then these extra dimensions could remain invisible even if they are very big. This possibility has caused a great deal of interest, because large extra dimensions might in principle be detectable by new sets of experiments, some of which might involve measuring the properties of gravity on a variety of scales, or else might involve looking for exotic phenomena at new particle accelerators. While it is indeed a long shot, it is nevertheless fascinating that our universe might be embedded in even an infinitely large space of higher dimensions, which might contain "nearby" four-dimensional or higher universes in which other galaxies, planets, and even aliens might exist, à la Star Trek. The only difference is that these universes will remain forever out of our reach, because our atoms, and the forces that bind them, might only operate in our four-dimensional space. So, while these ideas bring back shades of *Buckaroo Banzai Across the Eighth Dimension* (which interestingly has a Star Trek connection. I once read a quote from the movie neatly embossed on the lift on the *Deep Space Nine* set when I was visiting it with Michael Okuda), both he and Commander Riker would be safe from marauding aliens.

We also cannot rule out the possibility that there might exist microscopic or even macroscopic "bridges" to otherwise disconnected (or parallel) universes. Indeed, in general relativity, regions of very high curvature—inside a black hole, or in a wormhole—can be thought of as connecting otherwise disconnected and potentially very large regions of spacetime. I know of no reason to expect such phenomena outside black holes and wormholes, based on our present picture of the universe, but since we cannot rule them out, I suppose that Federation starships are free to keep finding them.

Anyons. In the *Next Generation* episode "The Next Phase," a transporter mix-up with a new Romulan cloaking device that puts matter "out of phase" with other matter causes Geordi LaForge and Ro Laren to vanish. They are presumed dead, and remain invisible and incommunicado until Data modifies an "anyon emitter" for another purpose and miraculously "dephases" them.

If the Star Trek writers had never heard of anyons, and I am willing to bet that they hadn't, their penchant for pulling apt names out of the air is truly eerie. Anyons are theoretical constructs proposed and named by my friend Frank Wilczek, a physicist at the Institute for Advanced Study in Princeton, and his collaborators. Incidentally, he also invented another particle—a dark matter candidate he called the axion, after a laundry detergent. "Axionic chips" also crop up in Star Trek, as part of an advanced machine's neural network. But I digress.

In the three-dimensional space in which we live, elementary particles are designated as fermions and bosons, depending on their spin. We associate with each variety of elementary particle a quantum number, which gives the value of its spin. This number can be an integer (0, 1, 2,. . .) or a half integer (1/2, 3/2, 5/2,. . .). Particles with integer spin are called bosons, and particles with half integer spin are called fermions. The quantum mechanical behavior of fermions and bosons is different: When two identical fermions are interchanged, the quantum mechanical wavefunction describing their properties is multiplied by minus 1, whereas in an interchange of bosons nothing happens to the wavefunction. Therefore, two fermions can never be in the same place, because if they were, interchanging them would leave the configuration identical but the wavefunction would have to be multiplied by minus 1, and the only thing that can be multiplied by minus 1 and remain the same is 0. Thus, the wavefunction must vanish. This is the origin of the famous Pauli exclusion principle originally applied to electrons—which states that two identical fermions cannot occupy the same quantum mechanical state.

In any case, it turns out that if one allows particles to move in only two dimensions—as the two-dimensional beings encountered by the *Enterprise* (see next item) are forced to do; or, more relevantly, as happens in the real world when atomic configurations in a crystal are arranged so that electrons, say, travel only on a two-dimensional plane—the standard quantum mechanical rules that apply in three-dimensional space are changed. Spin is no longer quantized, and particles can carry any value for this quantity. Hence, instead of fermi-ons or bos-ons, one can have any-ons. This was the origin of the name, and the idea that Wilczek and others have explored.

Back to the Star Trek writers: What I find amusing is that the number by which the wavefunction of particles is multiplied when the particles are interchanged is called a "phase." Fermion wavefunctions are multiplied by a phase of minus 1, while bosons are multiplied by a phase of 1 and hence remain the same. Anyons are multiplied by a combination of 1 and an imaginary number (imaginary numbers are the square roots of negative numbers), and hence in a real sense are "out of phase" with normal particles. So it seems fitting that an "anyon emitter" would change the phase of something, doesn't it?

Cosmic Strings. In the *Next Generation* episode "The Loss," the crew of the *Enterprise* encounters two-dimensional beings who have lost their way. These beings live on a "cosmic-string fragment." In the episode, this is described as an infinitesimally thin filament in space, with a very strong gravitational pull and vibrating with a characteristic set of "subspace" frequencies.

In fact, cosmic strings are objects proposed to have been created during a phase transition in the early universe. One of the world's experts on these theoretical objects is on the faculty at Case Western Reserve University, so I regularly hear about cosmic strings. Their properties would be similar in some respects to the object encountered by the *Enterprise.*

During a phase transition in materials—as when water boils, say, or freezes—the configuration of the material's constituent particles changes. When water freezes, it forms a crystalline structure. As crystals aligned in various directions grow, they can meet to form random lines, which create the patterns that look so pretty on a window in the winter. During a phase transition in the early universe, the configuration of matter, radiation, and empty space (which, I remind you, can carry energy) changes, too. Sometimes during these transitions, various regions of the universe relax into different configurations. As these configurations grow, they too can eventually meet—sometimes at a point, and sometimes along a line, marking a boundary between the regions. Energy becomes trapped in this boundary line, and it forms what we call a cosmic string.

We have no idea whether cosmic strings actually were created in the early universe, but if they were and lasted up to the present time they could produce some fascinating effects. They would be infinitesimally thin—thinner than a proton—yet the mass density they carry would be enormous, up to a million million tons per centimeter. They might form the seeds around which matter collapses to form galaxies, for example. They would also "vibrate," producing not subspace harmonics but gravitational waves. Indeed, we may well detect the gravitational wave signature of a cosmic string before we ever directly observe the string itself.

In the past decade, cosmic strings have taken on a new twist, with string theorists trying to get in on the action. It has been speculated that fundamental strings, the objects wiggling around in 10 or 11 dimensions, might yield 4-dimensional remnants that could be large macroscopic cosmic strings created in the very early universe, producing all of the effects that previously investigated cosmic strings might produce, and more. To date there remains no evidence for such, but it has rekindled interest in examining scenarios in which such objects may produce observational cosmological effects.

So much for the similarities with the Star Trek string. Now for the differences. Because of the way they are formed, cosmic strings cannot exist in fragments. They have to exist either in closed loops or as a single long string that winds its way through the universe. Moreover, in spite of their large mass density, cosmic strings exert no gravitational force on faraway objects. Only if a cosmic string moves past an object will the object experience a sudden gravitational force. These are subtle points, however; on the whole, the Star Trek writers have done pretty well by cosmic strings.

Quantum Measurements. There was a wonderful episode in the final season of *The Next Generation,* called "Parallels," in which Worf begins to jump between different "quantum realities." The episode touches, albeit incorrectly, on one of the most fascinating aspects of quantum mechanics—quantum measurement theory.

Since we live on a scale at which quantum mechanical phenomena are not directly observed, our entire intuitive physical picture of the universe is classical in character. When we discuss quantum mechanics, we generally use a classical language, so as to try and explain the quantum mechanical world in terms we understand. This approach, which is usually referred to as "the interpretation of quantum mechanics" and so fascinates some philosophers of science, is benighted; what we really should be discussing is "the interpretation of classical mechanics"—that is, how can the classical world we see—which is only an approximation of the underlying reality, which in turn is quantum mechanical in nature—be understood in terms of the proper quantum mechanical variables?

If we insist on interpreting quantum mechanical phenomena in terms of classical concepts, we will inevitably encounter phenomena that seem paradoxical, or impossible. This is as it should be. Classical mechanics cannot account properly for quantum mechanical phenomena, and so there is no reason that classical descriptions should make sense.

Having issued this caveat, I will describe the relevant issues in classical mechanics terms, because these are the only tools of language I have. While I have the proper mathematical terms to describe quantum mechanics, like all other physicists I have recourse only to a classical mental picture, because all my direct experience is classical.

As I alluded to in chapter 5, one of the most remarkable features of quantum mechanics is that objects observed to have some property cannot be said to have had that property the instant before the observation. The observation process can change the character of the physical system under consideration. The quantum mechanical wavefunction of a system describes completely the configuration of this system at any one time, and this wavefunction evolves according to deterministic laws of physics. However, what makes things seem so screwy is that this wavefunction can encompass two or more mutually exclusive configurations at the same time.

For example, if a particle is spinning clockwise, we say that its spin is "up." If it is spinning counterclockwise, we say that its spin is "down." Now, the quantum mechanical wavefunction of this particle can incorporate a sum with equal probabilities: spin up and spin down. If you measure the direction of the spin, you will measure *either* spin up *or* spin down. Once you have made the measurement, the wavefunction of the particle will from then on include only the component you measured the particle to have; if you measured spin up, you will go on measuring this same value for this particle.

This picture presents problems. How, you may ask, can the particle have had both spin up and spin down before the measurement? The correct answer is that it had neither. The configuration of its spin was indeterminate before the measurement.

The fact that the quantum mechanical wavefunction that describes objects does not correspond to unique values for observables is especially disturbing when one begins to think of living objects. There is a famous paradox called "Schrödinger's cat." (Erwin Schrödinger was one of the young Turks in their twenties who, early in the last century, helped uncover the laws of quantum mechanics. The equation describing the time evolution of the quantum mechanical wavefunction is known as Schrödinger's equation.) Imagine a box, inside of which is a cat. Inside the box, aimed at the cat, is a gun, which is hooked up to a radioactive source. The radioactive source has a certain quantum mechanical probability of decaying at any given time. When the source decays, the gun will fire and kill the cat. Is the wavefunction describing the cat, before I open the box, a linear superposition of a live cat and a dead cat? This seems absurd.

Similarly, our consciousness is always unique, never indeterminate. Is the act of consciousness a measurement? If so, then it could be said that at any instant there is a nonzero quantum mechanical probability for a number of different outcomes to occur, and our act of consciousness determines which outcome we experience. Reality then has an infinite number of branches. At every instant our consciousness determines which branch we inhabit, but an infinite number of other possibilities exist a priori.

This "many worlds" interpretation of quantum mechanics—which says that in some other branch of the quantum mechanical wavefunction Stephen Hawking is writing this book and I am writing the foreword—is apparently the basis for poor Worf's misery. Indeed, Data says as much during the episode. When Worf's ship traverses a "quantum fissure in spacetime," while simultaneously emitting a "subspace pulse," the barriers between quantum realities "break down," and Worf begins to jump from one branch of the wavefunction to another at random times, experiencing numerous alternative quantum realities. This can never happen, of course, because once a measurement has been made, the system, including the measuring apparatus (Worf, in this case), has changed. Once Worf has an experience, there is no going back . . . or perhaps I should say sideways. The experience itself is enough to fix reality. The very nature of quantum mechanics demands this.

There is one other feature of quantum mechanics touched upon in the same episode. The *Enterprise* crew are able to verify that Worf is from another "quantum reality" at one point by arguing that his "quantum signature at the atomic level" differs from anything in their world. According to Data, this signature is unique and cannot change due to any physical process. This is technobabble, of course; however, it does relate to something interesting about quantum mechanics. The entire set of all possible states of a system is called a Hilbert space, after David Hilbert, the famous German mathematician who, among other things, came very close to developing general relativity before Einstein. It sometimes happens that the Hilbert space breaks up into separate sectors, called "superselection sectors." In this case, no local physical process can move a system from one sector to another. Each sector is labeled by some quantity—for instance, the total electric charge of the system. If one wished to be poetic, one could say that this quantity provided a unique "quantum signature" for this sector, since all local quantum operations preserve the same sector, and the behavior of the operations and the observables they are associated with is determined by this quantity.

However, the different branches of the quantum mechanical wavefunction of a system must be in a single superselection sector, because any one of them is physically accessible in principle. So, unfortunately for Worf, even if he did violate the basic tenets of quantum mechanics by jumping from one branch to another, no external observable would be likely to exist to validate his story.

The whole point of the many-worlds interpretation of quantum mechanics (or any other interpretation of quantum mechanics, for that matter) is that you can never experience more than one world at a time. And thankfully there are other laws of physics that would prevent the appearance of millions of *Enterprises* from different realities, as happens at the end of the episode. Simple conservation of energy—a purely classical concept—is enough to forbid it.

Actually, there is one example I know of in the literature that suggests that different branches of the quantum wavefunction of the universe can be accessed, and it is an argument that has been espoused largely since the first edition of this book came out, in the context of possible time travel. Some have argued that one can get around the time travel paradoxes I mentioned earlier if, when traveling back in time, one travels to another branch of the quantum wavefunction. In this way, if one changes the future, it will not be the future on the branch that you started from, alleviating a lot of the problems. As nice as this sounds, however, I don't buy it, because I believe it violates various fundamental premises in quantum mechanics, including the conservation of probability. But nevertheless, I do find it amusing that in the series *Enterprise* that consummate time traveler, Crewman Daniels, constantly uses different exotic quantum devices associated with time travel. My favorite allusion is to the "quantum discriminator," which helps in targeting dates and times to travel to. Apparently by the thirty-first century physics has really progressed, and there is a quantum discriminator on every desk in high school!

On a much more realistic note, however, there has been a new development since the first edition of this book came out that truly exploits the weirdest aspect of quantum mechanics, namely that objects that are unobserved can exist in an indeterminate number of different states at the same time. This new development goes by the name "quantum computing," and it was first suggested perhaps by Richard Feynman, but has also been explored in depth originally by the physicists at IBM, who were so influential in promoting quantum teleportation, which in fact may be relevant to building realistic large-scale quantum computers, if such a thing ever becomes possible.

The idea behind quantum computing is relatively simple. Normal computers operate with bits, individual logical storage units which register either 1 or 0. Logical operations on these bits produce all of the computing phenomena that drive the modern world. Quantum computers would use "Qbits," or "qubits," instead of bits. A simple example of a qubit is the spinning particle I described earlier. Classically one could call spin-up 1, and spin-down 0. However, as I have described, before any observation, the quantum mechanical particle is actually spinning in all directions simultaneously. The idea behind quantum computing is to use this property to allow single qubits to be involved in many different computations at the same time! As long as one doesn't measure the individual qubits during the computational process (which would force them into a specific up or down state) one might therefore imagine a computer that would be orders of magnitude faster than current devices, because it could perform massively parallel computations. Indeed, it has been shown that this is possible in principle, and a quantum computer could perform some functions, like factoring large numbers, which normal computers would take longer than the age of the universe to do, instead in a finite, practical time. While this might not knock your socks off, you should be interested in this. This is because all banks currently keep your information secure by using codes based on large number factorization. Right now the codes cannot be broken by existing computers, but if quantum computers became practical, then this way of storing information securely would go out the window.

The big problem with quantum computers, however, is to make sure all of the parts maintain their "quantum coherence," even as information is passed around in the devices from qubit to qubit. This means, just as for quantum teleportation, the systems must remain isolated from all external noise and interactions. This is hard to do, especially for macroscopic sized systems, but we will wait and see. Perhaps quantum computers exist in *Star Trek*, which is why money doesn't seem to be important for much of the Federation. After all, if bank accounts and business transactions cannot be secure, what is the point of money?

Solitons. In the *Next Generation* episode "New Ground," the *Enterprise* assists in an experiment developed by Dr. Ja'Dor, of the planet Bilana III. Here a "soliton wave," a nondispersing wavefront of subspace distortion, is used to propel a test ship into warp speed without the need for warp drive. The system requires a planet at the far end of the voyage, which will deliver a scattering field to dissipate the wave. The experiment nearly results in a disaster, which is of course averted at the last instant.

Solitons are not an invention of the Star Trek writers. The term is short for "solitary waves" and in fact refers to a phenomenon originally observed in water waves by a Scottish engineer, John Scott Russell, in 1834. While conducting an unpaid study of the design of canal barges for the Union Canal Society of Edinburgh, he noticed something peculiar. In his own words:

> I was observing the motion of a boat which was rapidly drawn along a narrow channel by a pair of horses, when the boat suddenly stopped—Not so the mass of water in the channel which it had put in motion; it accumulated round the prow of the vessel in a state of violent agitation, then suddenly leaving it behind, rolled forward with great

velocity, assuming the form of a large solitary elevation, a rounded smooth and well defined heap of water, which continued its course along the channel apparently without change of form or diminution of speed. I followed it on horseback and overtook it still rolling on at a rate of some eight or nine miles an hour, preserving its original figure some thirty feet long and a foot to a foot and a half in height. Its height gradually diminished and after a chase of one or two miles I lost it in the windings of the channel. Such in the months of August 1834 was my first chance interview with that singular and beautiful phenomenon which I have called the Wave of Translation.[2]

Scott Russell later coined the words "solitary wave" to describe this marvel, and the term has persisted, even as solitons have cropped up in many different subfields of physics. More generally, solitons are nondissipative, classically extended, but finite-size objects that can propagate from point to point. In fact, for this reason the disasters that drive the plot in "New Ground" could not happen. First of all, the soliton would not "emit a great deal of radio interference." If it did, it would be dissipating its energy. For the same reason, it would not continue to gain energy or change frequency.

Normal waves are extended objects that tend to dissipate their energy as they travel. However, classical forces—resulting from some interaction throughout space, called a "field"—generally keep solitons intact, so that they can propagate without losing energy to the environment. Because they are self-contained energetic solutions of the equations describing motion, they behave, in principle, just like fundamental objects—like elementary particles. In fact, in certain mathematical models of the strong interaction holding quarks together, the proton could be viewed as a soliton, in which case we are all made of solitons! New fields have been proposed in elementary-particle physics which may coalesce into "soliton stars"—objects that are the size of stars but involve a single coherent field. Such objects have yet to be observed, but they may well exist.

Quasars. In the episode "The Pegasus"—wherein we learn about the Treaty of Algon, which forbade the Federation to use cloaking devices—we find Picard's *Enterprise* exploring the Mecoria Quasar. Earlier, in the original-series episode "The Galileo Seven," we learned that the original *Enterprise* had standing orders to investigate these objects whenever they might be encountered. But neither ship would in fact likely ever encounter a quasar while touring the outskirts of our galaxy. This is because quasars, the most energetic objects yet known in the universe (they radiate energies comparable to those of entire galaxies, yet they are so small that they are unresolvable by telescopes), are thought to be enormous black holes at the center of some galaxies, and to be literally swallowing up the central mass of their hosts. This is the only mechanism yet proposed that can explain the observed energies and size scales of quasars. As matter falls into a black hole, it radiates a great deal of energy (as it loses its potential gravitational energy). If million- or billion-solar-mass black holes exist at the centers of some galaxies, they can swallow whole star systems, which in turn will radiate the necessary energy to make up the quasar signal. For this reason, quasars are often part of what we call "active galactic nuclei." Also for this reason, you would not want to encounter one of these objects up close. The encounter would be fatal.

Neutrinos. Neutrinos are my favorite particles in nature, which is why I saved them for last. I have spent a fair fraction of my own research on these critters, because we know so little about them yet they promise to teach us much about the fundamental structure of matter and the nature of the universe.

Many times, in various Star Trek episodes, neutrinos are used or measured on starships. For example, elevated neutrino readings are usually read as objects traverse the Bajoran wormhole. We also learn in the episode "The Enemy" that Geordi LaForge's visor can detect neutrinos, when a neutrino beacon is sent to locate him so that he can be rescued from an inhospitable planet. A "neutrino field" is encountered in the episode "Power Play," and momentarily interferes with the attempt to transport some noncorporeal criminal life-forms aboard the *Enterprise.*

Neutrinos were first predicted to exist as the result of a puzzle related to the decay of neutrons. While neutrons are stable inside atomic nuclei, free neutrons are observed to decay, in an average time of about 10 minutes, into protons and electrons. The electric charge works out fine, because a neutron is electrically neutral, while a proton has a positive charge and an electron an equal and opposite negative charge. The mass of a proton plus an electron is almost as much as the mass of a neutron, so there is not much free energy left to produce other massive particles in the decay, in any case.

However, sometimes the proton and electron are observed to travel off in the same direction during the decay. This is impossible, because each emitted particle carries momentum. If the original neutron was at rest, it had zero momentum, so something else would have to be emitted in the decay to carry off momentum in the opposite direction.

Such a hypothetical particle was proposed by Wolfgang Pauli in the 1930s, and was named a "neutrino" (for "little neutron") by Enrico Fermi. He chose this name because Pauli's particle had to be electrically neutral, in order not to spoil the charge conservation in the decay, and had to have, at most, a very small mass, in order to be produced with the energy available after the proton and electron were emitted.

Because neutrinos are electrically neutral, and because they do not feel the strong force (which binds quarks and helps hold the nucleus together), they interact only very weakly with normal matter. Yet because neutrinos are produced in nuclear reactions, like those that power the Sun, they are everywhere. Six hundred billion neutrinos pierce every square centimeter of your body every second of every day, coming from the Sun—an inexorable onslaught that has even inspired a poem by John Updike. You don't notice this neutrino siege, because the neutrinos pass right through your body without a trace. On average, these solar neutrinos could go through 10,000 light-years of material before interacting with any of it.

If this is the case, then how can we be sure that neutrinos exist other than in theory, you may ask? Well, the wonderful thing about quantum mechanics is that it yields probabilities. That is why I wrote "on average" in the above paragraph. While most neutrinos will travel 10,000 light-years through matter without interacting with anything, if one has enough neutrinos and a big enough target, one can get lucky.

This principle was first put to use in 1956 by Frederick Reines and Clyde Cowan, who put a several-ton target near a nuclear reactor and indeed observed a few events. This empirical discovery of the neutrino (actually, the antineutrino) occurred more than 20 years after it was posited, and well after most physicists had accepted its existence. Almost 40 years later, in the year the first edition of this book appeared, Reines won the Nobel Prize for his contribution to the experimental discovery of the neutrino.

Nowadays we use much larger detectors. The first observation of solar neutrinos was made in the 1960s, by Ray Davis and collaborators, using 100,000 gallons of cleaning fluid in a tank underground at the Homestake Gold Mine in South Dakota. Each day, on average, one neutrino from the Sun would interact with an atom of chlorine and turn it into an atom of argon. It is a tribute to these experimenters that they could detect nuclear alchemy at such a small rate. It turns out that the rate that their detector and all subsequent solar-neutrino detectors measured is different from the predicted rate. This "solar neutrino puzzle," as it is called, is now understood to signal the need for new fundamental physics associated with neutrinos. This discovery by Ray Davis in the early 1960s was honored by a Nobel Prize in Physics, in 2002.

The biggest neutrino detector in the world was built in the Kamiokande mine in Japan. Containing over 50,000 tons of water, it has been the successor to a 5000-ton detector, which was one of two neutrino detectors to see a handful of neutrinos from a 1987 supernova in the Large Magellanic Cloud, more than 150,000 light-years away! The Nobel Prize in 2002 was actually shared with Davis by Masatoshi Koshiba, who directed the construction of these detectors in Japan.

Which brings me back to where I began. Neutrinos are one of the new tools physicists are using to open windows on the universe. By exploiting every possible kind of elementary-particle detection along with our conventional electromagnetic detectors, we may well uncover the secrets of the galaxy long before we are able to venture out and explore it. Of course, if it were possible to invent a neutrino detector the size of Geordi's visor, that would be a great help!

10

Impossibilities

The Undiscoverable Country

GEORDI: "Suddenly it's like the laws of physics went right out the window."
Q: "And why shouldn't they? They're so inconvenient!"

—In "True Q"

"Bones, I want the impossible checked out too."

—Kirk to McCoy, in "The Naked Time"

"What you're describing is . . . nonexistence!"

—Kirk to Spock, in "The Alternative Factor"

Any sensible trekker-physicist recognizes that Star Trek must be taken with a rather large grain of salt. Nevertheless, there are times when for one reason or another the Star Trek writers cross the boundaries from the merely vague or implausible to the utterly impossible. While finding even obscure technical flaws with each episode is a universal trekker pastime, it is not the subtle errors that physicists and physics students seem to relish catching. It is the really big ones that are most talked about over lunch and at coffee breaks during professional meetings.

To be fair, sometimes a sweet piece of physics in the series—even a minor moment—can trigger a morning-after discussion at coffee time. Indeed, I remember vividly the day when a former graduate student of mine at Yale—Martin White, who is now a professor at the University of California, Berkeley—came into my office fresh from seeing *Star Trek VI: The Undiscovered Country*. I had thought we were going to talk about gravitational waves from the very early universe. But instead Martin started raving about one particular scene from the movie—a scene that lasted all of about 15 seconds. Two helmeted assassins board Chancellor Gorkon's vessel—which has been disabled by photon torpedoes fired from the *Enterprise* and is thus in zero gravity conditions—and shoot everyone in sight, including Gorkon. What impressed Martin and, to my surprise, a number of other physics students and faculty I discussed the movie with, was that the drops of blood flying about the ship were spherical. On Earth, all drops of liquid are tear-shaped, because of the relentless pull of gravity. In a region devoid of gravity, like Gorkon's ship, even tears would be spherical. Physicists know this but seldom have the opportunity to see it. So by getting this simple fact perfectly right, the Star Trek special effects people made a lot of physics types happy. It doesn't take that much. Actually, the Star Trek writers did a nice reprise of this same phenomenon in the later series *Enterprise*, where Captain Archer is shown taking a shower with water (sonic showers were apparently introduced later), and the ship experiences a malfunction of its artificial gravity system, and nice spherical globules of water start floating around, as does Archer himself. I don't know if the writers did this because they knew from the first edition of this book that we physicists liked the original scene so much, but I like to think that might be possible . . .

But the mistakes also keep us going. In fact, what may be the most memorable Star Trek mistake mentioned by a physicist doesn't involve physics at all. It was reported to me by the particle physicist (and science writer) Steven Weinberg, who won the Nobel Prize for helping develop what is now called the Standard Model of elementary particle interactions. As I knew that he keeps the TV on while doing intricate calculations, I wrote to him and asked for his Star Trek memories. Weinberg replied that "the main mistake made on Star Trek is to split an infinitive every damn time: To boldly go . . . !"

More often than not, though, it is the physics errors that get the attention of physicists. I think this is because these mistakes validate the perception of many physicists that physics is far removed from popular culture—not to mention the superior feeling it gives us to joke about the English majors who write the show. It is impossible to imagine that a major motion picture would somehow have Napoleon speaking German instead of French, or date the signing of the Declaration of Independence in the nineteenth century. And so when a physics mistake of comparable magnitude manages to creep into what is after all supposed to be a scientifically oriented series, physicists like to pounce. I was surprised to find out how many of my distinguished colleagues—from Kip Thorne to Weinberg to Sheldon Glashow, not to mention Stephen Hawking, perhaps the most famous physicist trekker of all—have watched the Star Trek series. Here is a list of my favorite blunders, gleaned from discussions with these and other physicists and e-mail from techni-trekkers. I have made an effort here to focus mostly (but not exclusively) on blunders of "down-to-Earth physics." Thus, for example, I don't address such popular complaints as "Why does the starlight spread out whenever warp speed is engaged?" and the like. Similarly, I ignore here the technobabble—the indiscriminate use of scientific and pseudoscientific terminology used during each episode to give the flavor of futuristic technology. Finally, I have tried for the most part to choose examples I haven't discussed before.

"In Space, No One Can Hear You Scream." The promo for *Alien* got it right, but Star Trek usually doesn't. Sound waves *DO NOT* travel in empty space! Yet when a space station orbiting the planet Tanuga IV blows up, from our vantage point aboard the *Enterprise* we hear it as well as see it. What's worse, we hear it *at the same time* as we see it. Even if sound

waves could travel in space, which they can't, the speed of a pressure wave such as sound is generally orders of magnitude smaller than the speed of light. You don't have to go farther than a local football game to discover that you see things before you hear them.

A famous experiment in high school physics involves putting an electric buzzer in a bell jar, a glass container from which the air can be removed by a pump. When the air is removed, the sound of the buzzer disappears. As early as the seventeenth century, it was recognized that sound needed some medium to travel in. In a vacuum, such as exists inside the bell jar, there is nothing to carry the sound waves, so you don't hear the buzzer inside. To be more specific, sound is a pressure wave, or disturbance, which moves as regions where the pressure is higher or lower than the average pressure propagate through a medium. Take away the medium, and there is no pressure to have a disturbance in. Incidentally, the bell jar example was at the origin of a mystery I discussed earlier, which was very important in the history of physics. For while you cannot hear the buzzer, you *can still see* it! Hence, if light is supposed to be some sort of wave, what medium does it travel in which isn't removed when you remove the air? This was one of the prime justifications for the postulation of the aether.

I had never taken much notice of the sound or lack of it in space in the series. However, after Steven Weinberg and several others mentioned that they remembered sound associated with Star Trek explosions, I checked the episode I had just watched—"A Matter of Perspective," the one in which the Tanuga IV space station explodes. Sure enough, *kaboom!* The same thing happened in the next episode I watched (when a shuttle that was carrying stolen trilithium crystals away from the *Enterprise* blew up with a loud bang near the planet Arkaria). I next went to the Star Trek movie *Generations.* There, even a bottle of champagne makes noise when it explodes in space.

In fact, a physics colleague, Mark Srednicki of U.C. Santa Barbara, brought to my attention a much greater gaffe in one episode, in which sound waves are used as a weapon against an orbiting ship. As if that weren't bad enough, the sound waves are said to reach "18 to the 12th power decibels." What makes this particularly grate on the ear of a physicist is that the decibel scale is a logarithmic scale, like the Richter scale. This means that the number of decibels already represents a power of 10, and they are normalized so that 20 decibels is 10 times louder than 10 decibels, and 30 decibels is 10 times louder again. Thus, 18 to the 12th power decibels would be $10^{18^{12}}$, or 1 followed by 11,568,313,814,300 zeroes times louder than a jet plane!

Faster Than a Speeding Phaser. While faster-than-light warp travel is something we must live with in Star Trek, such a possibility relies on all the subtleties of general relativity and exotic new forms of matter, as I have described. But for normal objects doing everyday kinds of things, light speed is and always will be the ultimate barrier. Sometimes this simple fact is forgotten. In a wild episode called "Wink of an Eye," Kirk is tricked by the Scalosians into drinking a potion that speeds up his actions by a huge factor to the Scalosian level, so that he can become a mate for their queen, Deela. The Scalosians live a hyperaccelerated existence and cannot be sensed by the *Enterprise*'s crew. Before bedding the queen, Kirk first tries to shoot her with his phaser. However, since she can move in the wink of an eye by normal human standards, she moves out of the way before the beam can hit her. Now, what is wrong with this picture? The answer is, Everything!

What has been noticed by some trekkers is that the accelerated existence required for Deela to move significantly in the time it would take a phaser beam to move at the speed of light across the room would make the rest of the episode impossible. Light speed is 300 million meters per second. Deela is about a meter or so away from Kirk when he fires, implying a light travel time of about 1/300 millionth of a second. For this time to appear to take a second or so for her, the Scalosian clock must be faster by a factor of 300 million. However, if this is so, 300 million Scalosian seconds take 1 second in normal *Enterprise* time. Unfortunately, 300 million seconds is about 10 years.

OK, let's forgive the Star Trek writers this lapse. Nevertheless, there is a much bigger problem, which is impossible to solve and which several physicists I know have leapt upon. Phasers are, we are told, directed energy weapons, so that the phaser beam travels at the speed of light. Sorry, but there is no way out of this. If phasers are pure energy and not particle beams, as the Star Trek technical manual states, the beams must move at the speed of light. No matter how fast one moves, even if one is sped up by a factor of 300 million, one can never move out of the way of an oncoming phaser beam. Why? Because in order to know it is coming, you have to first see the gun being fired. But the light that allows you to see this travels at the same speed as the beam. Put simply, it is impossible to know it is going to hit you until it hits you! As long as phaser beams are energy beams, there is no escape. A similar problem involving the attempt to beat a phaser beam is found in the *Voyager* episode "The Phage."

Sometimes, however, it is the Star Trek critics who make the mistakes. I was told that I should take note of an error in *Generations* in which a star shining down on a planet is made to disappear and at the same instant the planet darkens. This of course is impossible, because it takes light a finite time to travel from the star to the planet. Thus, when I turn off the light from a star, the planet will not know it for some time. However, in *Generations,* the whole process is seen from the surface of the planet. When viewed from the planet, the minute the star is seen to implode, the planet's surface should indeed get dark. This is because both the information that the star has imploded and the lack of light will arrive at the planet at the same time. Both will be delayed, but they will be coincident!

Though the writers got this right, they blew it by collapsing the delay to an unreasonably short time. We are told that the probe that will destroy the star will take only 11 seconds to reach it after launch from the planet's surface. The probe is traveling at sublight speeds—as we can ascertain because it takes much less than twice that time after the probe is launched for those on the planet to see the star begin to implode, which indicates that the light must have taken fewer than 11 seconds to make the return journey. The Earth, by comparison, is 8 light-minutes from our Sun, as I have noted. If the Sun exploded now, it would take 8 minutes for us to know about it. I find it hard to believe that the Class M planet in *Generations* could exist at a distance of 10 light-seconds from a hydrogen-burning star like our Sun. This distance is about 5 times the size of the Sun—far too close for comfort.

If the Plot Isn't Cracked, Maybe the Event Horizon Is. While I said I wouldn't dwell on technobabble, I can't help mentioning that the *Voyager* series wins in that department hands down. Every piece of jargon known to modern physics is thrown in as the *Voyager* tries to head home, traveling in time with the regularity of a commuter train. However, physics terms usually *mean* something, so that when you use them as a plot device you are bound to screw up every now and then. I mentioned in chapter 3 that the "crack" in the event horizon that saves the day for the *Voyager* (in the feckless "Phage" episode) sounds particularly ludicrous to physicists. A "crack" in an event horizon is like removing one end of a circle, or like being a little bit pregnant. It doesn't mean anything. The event horizon around a black hole is not a physical entity, but rather a location inside of which all trajectories remain inside the hole. It is a property of curved space that the trajectory of anything, including light, will bend back toward the hole once you are inside a certain radius. Either the event horizon exists, in which case a black hole exists, or it doesn't. There is no middle ground big enough to slip a needle through, much less the *Voyager.*

How Solid a Guy Is the Doctor? I must admit that the technological twist I like the most in the *Voyager* series was the holographic doctor. There is a wonderful scene in which a patient asks the doctor how he can be solid if he is only a hologram. This is a good question. The doctor answers by turning off a "magnetic confinement beam" to show that without it he is as noncorporeal as a mirage. He then orders the beam turned back on, so that he can slap the poor patient around. It's a great moment, but unfortunately it's also an impossible one. As I described in chapter 6, magnetic confinement works wonders for charged particles, which experience a force in a constant magnetic field that causes them to move in circular orbits. However, light is not charged. It experiences no force in a magnetic field. Since a hologram is no more than a light image, neither is the doctor.

Which Is More Sensitive, Your Hands or Your Butt? Or, to Interphase, or Not to Interphase. Star Trek has on occasion committed what I call the infamous *Ghost* error. I refer to a 1990 movie by this name in which the main character, a ghost, walks through walls and cannot lift objects because his hand passes through them. However, miraculously, whenever he sits on a chair or a couch, his butt manages to stay put. Similarly, the ground seems pretty firm beneath his feet. In the last chapter, I described how Geordi LaForge and Ro Laren were rendered "out of phase" with normal matter by a Romulan "interphase generator." They discovered to their surprise that they were invisible and could walk through people and walls—leading Ro, at least, to believe that she was dead (perhaps she saw a replay of *Ghost* at some old movie house in her youth). Yet Geordi and Ro could stand on the floor and sit on chairs with impunity. Matter is matter, and chairs and floors are no different from walls, and as far as I know feet and butts are no more or less solid than hands.

Incidentally, there is another fatal flaw associated with this particular episode which also destroys the consistency of a number of other Star Trek dramas. In physics, two things that both interact with something else will always be able to interact with each other. This leads us full circle back to Newton's First Law. If I exert a force on you, you exert an equal and opposite force on me. Thus, if Geordi and Ro could observe the *Enterprise* from their new "phase," they could interact with light, an electromagnetic wave. By Newton's Law if nothing else, they in turn should have been visible. Glass is invisible precisely because it does not absorb visible light. In order to see—that is, to sense light—you have to absorb it. By absorbing light, you must disturb it. If you disturb light, you must be visible to someone else. The same goes for the invisible interphase insects that invaded the *Enterprise* by clinging to the bodies of the crew, in the *Next Generation* episode "Phantasms." The force that allows them to rest on normal matter without going through it is nothing other than electromagnetism—the electrostatic repulsion between the charged particles making up the atoms in one body and the atoms in another body. Once you interact electromagnetically, you are part of our world. There is no such thing as a free lunch.

Actually, after the first edition of this book appeared and I discussed this blooper at a public lecture, a six-year-old boy (I wish I knew his name, as he might now be in college studying physics!) asked me a very interesting question. If Geordi and Ro were out of phase, how could they breath? Wouldn't the air go right through their bodies? Good point, that I hadn't thought of—out of the mouth of a child!

Sweeping Out the Baby with the Bathwater. In the *Next Generation* episode "Starship Mine," the *Enterprise* docks at the Remmler Array to have a "baryon sweep." It seems that these particles build up on starship superstructures as a result of long-term travel at warp speed, and must be removed. During the sweep, the crew must evacuate, because the removal beam is lethal to living tissue. Well, it certainly would be! The only stable baryons are (1) protons and (2) neutrons in atomic nuclei. Since these particles make up everything we see, ridding the *Enterprise* of them wouldn't leave much of it for future episodes.

How Cold Is Cold? The favorite Star Trek gaffe of my colleague and fellow Star Trek aficionado Chuck Rosenblatt involves an object's being frozen to a temperature of –295° Celsius. This is a very exciting discovery, because on the Celsius scale, absolute zero is –273°. Absolute zero, as its name implies, is the lowest temperature anything can potentially attain, because it is defined as the temperature at which all molecular and atomic motions, vibrations, and rotations cease. Though it is impossible to achieve this theoretical zero temperature, atomic systems have been cooled to within a millionth of a degree above it (and as of this writing have just been cooled to 2 billionths of a degree above absolute zero). Since temperature is associated with molecular and atomic motion, you can never get less than no motion at all; hence, even 400 years from now, absolute zero will still be absolute.

I Have Seen the Light! I am embarrassed to say that this obvious error, which I should have caught myself, was in fact pointed out to me by a first-year physics student, Ryan Smith, when I was lecturing to his class and mentioned that I was writing this book. Whenever the *Enterprise* shoots a phaser beam, we see it. But of course this is impossible unless the phaser itself emits light in all directions. Light is not visible unless it reflects off something. If you have ever been to a lecture given with the help of a laser pointer—generally, these are helium-neon red lasers—you may recall that you see only the spot where the beam hits the screen, and not anything in between. The only way to make the whole beam visible is to make the room dusty, by clapping chalkboard erasers together, or something like that. (You should try this sometime; the light show is really quite spectacular.) Laser light shows are created by bouncing the laser light off either smoke or water. Thus, unless empty space is particularly dusty, we shouldn't see the phaser beam except where it hits.

Astronomers Get Picky. Perhaps it is not surprising to find that the physics errors various people find in the series are often closely related to their own areas of interest. As I polled people for examples, I invariably got responses that bore a correlation to the specific occupations of those who volunteered the information. I received several responses by e-mail from astronomer-trekkers who reacted to several subtle Star Trek errors. One astronomy student turned a valiant effort by the Star Trek writers to use a piece of real astronomy into an error. The energy-eating life-form in "Galaxy's Child" is an infant space creature, who mistakes the *Enterprise* for its mother and begins draining its energy. Just in the nick of time LaForge comes up with a way to get the baby to let go. The baby is attracted to the radiation the *Enterprise* is emitting, at a 21-cm wavelength. By changing the frequency of the emission, the crew "spoils the milk," and the baby lets go. What makes this episode interesting, and at the same time incorrect, is that the writers picked up on a fact I mentioned in chapter 8—namely, the 21-cm radiation is a universal frequency emitted by hydrogen, which astronomers use to map out interstellar gas. However, the writers interpreted this to mean that everything radiates at 21 cm, including the *Enterprise.* In fact, the atomic transition in hydrogen responsible for this radiation is extremely rare, so that a particular atom in interstellar space might produce such radiation on average only once every 400 years. However, because the universe is filled with hydrogen, the 21-cm signal is strong enough to detect on Earth. So, in this case, I would give the writers A for effort and reduce this grade to B+ for the misinterpretation—but I am known as an easy grader.

A NASA scientist pointed out an error I had missed and which you might expect someone working for NASA to recognize. It is generally standard starship procedure to move into geosynchronous orbit around planets—that is, the orbital period of the ship is the same as that of the planet. Thus the ship should remain above the same place on the planet's surface, just as geosynchronous weather satellites do on Earth. Nevertheless, when the *Enterprise* is shown orbiting a planet it is usually moving against the background of the planet's surface. And indeed, if it is not in a geosynchronous orbit, then you run into considerable beaming-up problems.

Some new contributions since the first edition. Of course there have been a host of bloopers in the years since this book first appeared, both in the intervening series and movies and also ones from the earlier series that others have reminded me of. I cannot do justice here to all of them, so I have picked four of the less obvious ones that came to me from different sources: a physicist, a fan, a trekker-editor, and a crew member:

> The physicist: My colleague Chuck Rosenblatt told me of an amusing mistake in the original series, in which Kirk orders the crew to increase the speed of the ship by "1 to the 20th power!" What is this? It is 1 x 1 x 1 . . . 20 times, but that still equals 1.
>
> The fan: Jeana Park is a fan who has written to me with many questions about the series, notably *Enterprise.* She pointed out a very simple physics blooper (my favorite kind) involving an Earth freighter attacked by Nausicaan pirates. Early in the episode the captain and the first officer are throwing a football in the cargo hold. They just nudge the ball and it floats back and forth between the two of them, indicating a close to zero gravity environment. But they walk around completely normally. Gravity affects things equally, so unless they are wearing gravity boots, like the assassins in *Star Trek VI*, they should float around along with the football.
>
> The editor: My editor for the new edition. William Frucht reminded me of an *Enterprise* blooper that Ms.

Park also asked about. It involves the crew's visit to a rogue planet—that is, a planet that has escaped its solar system and thus does not have a star. Why the planet should be warm and have life without a star is implausible to imagine, but not impossible; it could perhaps have huge amounts of internal radioactivity warming it, and the process that removed it from its host star may have been gentle enough not to strip the atmosphere. However, much of the action takes place in a lush jungle with trees. But if there is no star, there is no light. With no light there are no photons. Without photons, there is no photosynthesis. Thus, no trees with green leaves, please.

The crew member: At a fortieth anniversary Star Trek event I spent part of a pleasant evening drinking with Dr. Phlox (John Billingsley), who was trying to explain to me the weird mating relationships of Denobulans, the species his character portrayed. Each Denobulan woman has three husbands, and each man has three wives. This is fine, if three women, for example, are allowed to be wives of the same two men. But if not, which was Billingsley's perception (and he should know!), then there is a logistical nightmare, which we continued to try to sort out, unsuccessfully, over many drinks. If every woman has to find three different men, who must be married to two other different women, the number of interrelationships quickly grows, and in fact cannot close on itself with a finite number of people. Of course, since Denobulans can be intimate with anyone they choose in any case, the possibilities probably seem infinite in any case.

Those Darned Neutrinos. I can't help but close by bringing up neutrinos again. And since I have skipped lightly over *Deep Space Nine* in this book perhaps it is fair to finish with a blooper from this series—one I was told about by David Brahm, another physicist trekker. It seems that Quark has gotten hold of a machine that alters the laws of probability in its vicinity. One can imagine how useful this would be at his gambling tables, providing the kind of unfair advantage that a Ferengi couldn't resist. This ruse is discovered, however, by Dax, who happens to analyze the neutrino flux through the space station. To her surprise, she finds that all the neutrinos are coming through left-handed—that is, all spinning in one direction relative to their motion. Something must be wrong! The neutrinos that spin in the opposite direction seem to be missing!

Unfortunately, of all the phenomena the Star Trek writers could have chosen to uncover Quark's shenanigans, they managed to pick one that is actually true. As far as we know, neutrinos *are* only left-handed! They are the only known particles in nature that apparently can exist in only one spin state. If Dax's analysis had yielded this information, she would have every reason to believe that all was as it should be.

What makes this example so poignant, as far as I am concerned, is exactly what makes the physics of Star Trek so interesting: sometimes truth is indeed stranger than fiction.

EPILOGUE

Well, that's it for blunders and for physics. In the first edition I made a joke, picked up by the *New York Times,* suggesting that if enough people wrote in I would consider a sequel, titled *The Physics of Star Trek II: The Wrath of Krauss.*

I hope by now it is clear that was indeed just a joke, and there will be no sequel. Of course, if the *Star Trek* saga continues, I may continue to try to keep both the physics and the references to the series up to date . . . at least as long as my publishers can induce me to produce new editions.

The point of finishing this book with a chapter on physics blunders was not to castigate the Star Trek writers unduly. It is rather to illustrate that there are many ways of enjoying the series. As long as Star Trek continues to remain on the air, I am sure that ever-new physics faux pas will give trekkers of all ilks, from high school students to university professors, something to look forward to talking about the morning after. And it offers a challenge to the writers and producers to try to keep up with the expanding world of physics.

So I will instead close this book where I began—not with the mistakes but with the possibilities. Our culture has been as surely shaped by the miracles of modern physics—and here I include Galileo and Newton among the moderns—as it has by any other human intellectual endeavor. And while it is an unfortunate modern misconception that science is somehow divorced from culture, it is, in fact, a vital part of what makes up our civilization. Our explorations of the universe represent some of the most remarkable discoveries of the human intellect, and it is a pity that they are not shared among as broad an audience as enjoys the inspirations of great literature, or painting, or music.

By emphasizing the potential role of science in the development of the human species, Star Trek whimsically displays the powerful connection between science and culture. While I have argued at times that the science of the twenty-third century may bear very little resemblance to anything the imaginations of the Star Trek writers have come up with, nevertheless I expect that this science may be even more remarkable. In any case I am convinced that the physics of today and tomorrow will as surely determine the character of our future as the physics of Newton and Galileo colors our present existence. I suppose I am a scientist in part because of my faith in the potential of our species to continue to uncover hidden wonders in the universe. And this is after all the spirit animating the Star Trek series. Perhaps Gene Roddenberry should have the last word. As he said on the twenty-fifth anniversary of the Star Trek series, one year before his death: "The human race is a remarkable creature, one with great potential, and I hope that Star Trek has helped to show us what we can be if we believe in ourselves and our abilities."

NOTES

CHAPTER 1: NEWTON ANTES

1. Michael Okuda, Denise Okuda, and Debbie Mirak, *The Star Trek Encyclopedia* (New York: Pocket Books, 1994).

2. Rick Sternbach and Michael Okuda, *Star Trek: The Next Generation—Technical Manual* (New York: Pocket Books, 1991).

CHAPTER 2: EINSTEIN RAISES

1. Quoted in Paul Schilpp, ed., *Albert Einstein: Philosopher-Scientist* (New York: Tudor, 1957).

2. Rick Sternbach and Michael Okuda, *Star Trek: The Next Generation—Technical Manual* (New York: Pocket Books, 1991).

3. Ibid.

CHAPTER 3: HAWKING SHOWS HIS HAND

1. Michael Okuda, Denise Okuda, and Debbie Mirak, *The Star Trek Encyclopedia* (New York: Pocket Books, 1994).

CHAPTER 8: THE SEARCH FOR SPOCK

1. Review by Philip Morrison, in *Scientific American,* November 1994, of Hölldobler and Wilson, *Journey to the Ants: A Story of Scientific Explorations* (Cambridge, MA: Harvard University Press, 1994).

2. Francis Crick, *Life Itself* (New York: Simon & Schuster, 1981).

3. Bernard M. Oliver, "The Search for Extraterrestrial Life," *Engineering and Science,* December 1974.

CHAPTER 9: THE MENAGERIE OF POSSIBILITIES

1. For a cogent review of this subject, I suggest my own book *The Fifth Essence: The Search for Dark Matter in the Universe* (New York: Basic Books, 1989).

2. John Scott Russell, *Report of the 14th Meeting of the British Association for the Advancement of Science* (London: John Murray, 1844).

ACKNOWLEDGMENTS

I owe a great debt to many individuals who helped make this book possible. First, I am grateful to colleagues in the physics community, who responded unfailingly to requests for assistance. I thank in particular Stephen Hawking for readily agreeing to write the foreword, and Steven Weinberg, Sheldon Glashow, and Kip Thorne for providing me with their Star Trek memories. John Peoples, director of the Fermi National Accelerator Laboratory, made members of his staff available to help me report on antimatter production and storage at Fermilab. I particularly thank Judy Jackson of the Fermilab public relations office for assistance and photographs, and my Case Western Reserve University colleague Cyrus Taylor, who is currently performing an experiment at Fermilab, for answering various technical questions. Paul Horowitz of Harvard University responded to my request for information on the SETI and META programs that he has led by sending me within a day a treasure trove of information on the search for extraterrestrial intelligence and photographs of the projects. George Smoot provided the wonderful COBE photograph of our galaxy in the text, and Philip Taylor pointed me to a reference on solitons.

A number of trekker-physicists freely offered their insights about Star Trek physics. In particular, I am grateful to Mark Srednicki, Martin White, Chuck Rosenblatt, and David Brahm for pointing out useful examples from the series. I also want to thank the trekkers who responded to my e-mail request to the Star Trek bulletin boards for favorite pieces of physics and favorite bloopers, notably Scott Speck, "Westy" at NASA, T. J. Goldstein, Denys Proteau, and J. Dilday, for either reinforcing my own choices or suggesting other useful examples. I also want to thank various students at Case Western Reserve for volunteering information, especially Ryan Smith.

Other trekkers made important contributions. I want to thank Anna Fortunato for reading and commenting on early drafts of the manuscript and making many useful suggestions. Mark Landau at HarperCollins also supplied me with useful feedback. Jeffrey Robbins, at the time an editor at Oxford University Press, was gracious enough to provide me with an important reference on the warp drive. My uncle Herb Title, an avid trekker, read through the manuscript, as did my physics research associate Peter Kernan. Both provided useful comments. I also relied on my wife, Kate, for input on various parts of the manuscript.

I am indebted to Greg Sweeney and Janelle Keberle for loaning me their complete indexed collection of Star Trek videotapes, which I had at my disposal for four months during the writing of this book. These were essential for me, and they were used constantly to check information and verify plots. I thank them for entrusting me with their collection.

I want to offer very special thanks to my editor at Basic Books, Susan Rabiner, without whom this project would never have happened. She is the one who finally convinced me to tackle it, and she went out on every limb possible to help promote the project within Basic and HarperCollins. In this regard I also thank Kermit Hummel, former president of Basic Books, for his support and enthusiasm. The final form of this book also depended crucially on the wisdom and insights of Sara Lippincott, my line editor. The many hours we spent at the fax machine and on the telephone are reflected, I believe, in a substantially improved manuscript.

Finally, I want to thank the dean, faculty, staff, and students of the College of Arts and Sciences and the Physics Department at Case Western Reserve University for their support and, too often, their indulgence while this work was being completed. The atmosphere of collegiality and excitement they helped to foster kept me invigorated when I really needed to be.

As always, my family has supported my efforts however they could. Kate and my daughter, Lilli, even consented to watch Star Trek episodes with me late into the night on numerous occasions when they may have preferred to sleep.

Finally, for the revised edition of this book, I want to thank all of those individuals who have written to me over the intervening years with questions, suggestions, and complaints, as well as my physics colleagues, including a number of new Nobel laureates such as Gerard 't Hooft, who have motivated and provoked me to keep thinking about how the real universe and the Star Trek universe may intersect. I hope you may see a little of your own curiosity reflected in the new material you have found here.

INDEX

ABOUT THE AUTHOR

Lawrence M. Krauss is Ambrose Swasey Professor of Physics, Professor of Astronomy, and Director of the Center for Education and Research in Cosmology and Astrophysics at Case Western Reserve University. The author of seven popular books including *Fear of Physics* and the award winners *Atom* and *Hiding in the Mirror*, Krauss is also a regular radio commentator and essayist for newspapers such as *The New York Times* and appears regularly on television. He is one of the few well-known scientists today described by such magazines as *Scientific American* as a public intellectual, and with activities including performing with the Cleveland Orchestra and judging at the Sundance Film Festival, he has also crossed the chasm between science and popular culture. He is a highly regarded international leader in cosmology and astrophysics and is the author of over 200 papers. Krauss has won numerous international awards for his research accomplishments and writing (including the highest awards of the American Physical Society, the American Association of Physics Teachers, and the American Institute of Physics). He is a Fellow of the American Physical Society and the American Association for the Advancement of Science. He has been particularly active leading the effort by scientists to defend the teaching of science in public schools.